SPATIAL REGRESSION ANALYSIS USING EIGENVECTOR SPATIAL FILTERING

Spatial Econometrics and Spatial Statistics

SPATIAL REGRESSION ANALYSIS USING EIGENVECTOR SPATIAL FILTERING

DANIEL A. GRIFFITH
YONGWAN CHUN
BIN LI

Foreword by
PIERRE LEGENDRE

Series Editor
GIUSEPPE ARBIA

ELSEVIER

ACADEMIC PRESS
An imprint of Elsevier

Academic Press is an imprint of Elsevier
125 London Wall, London EC2Y 5AS, United Kingdom
525 B Street, Suite 1650, San Diego, CA 92101, United States
50 Hampshire Street, 5th Floor, Cambridge, MA 02139, United States
The Boulevard, Langford Lane, Kidlington, Oxford OX5 1GB, United Kingdom

Notices
Knowledge and best practice in this field are constantly changing. As new research and experience broaden our understanding, changes in research methods, professional practices, or medical treatment may become necessary.

Practitioners and researchers must always rely on their own experience and knowledge in evaluating and using any information, methods, compounds, or experiments described herein. In using such information or methods they should be mindful of their own safety and the safety of others, including parties for whom they have a professional responsibility.

To the fullest extent of the law, neither the Publisher nor the authors, contributors, or editors, assume any liability for any injury and/or damage to persons or property as a matter of products liability, negligence or otherwise, or from any use or operation of any methods, products, instructions, or ideas contained in the material herein.

Library of Congress Cataloging-in-Publication Data
A catalog record for this book is available from the Library of Congress

British Library Cataloguing-in-Publication Data
A catalogue record for this book is available from the British Library

ISBN 978-0-12-815043-6

For information on all Academic Press publications
visit our website at https://www.elsevier.com/books-and-journals

Publisher: Candice Janco
Acquisition Editor: J. Scott Bentley
Editorial Project Manager: Kelsey Connors
Production Project Manager: Maria Bernard
Cover Designer: Matthew Limbert

Typeset by SPi Global, India

This book is dedicated to the memory of Dr. Ruth I. Shirey, whose mentoring and support enabled Dan Griffith to co-author it; Yongwan Chun's family, Hyunju Lee and Christopher Chun; and Bin Li's wife, Maiokun.

Contents

Foreword

Geographers have long understood that natural phenomena contain spatial structures. Tobler's First Law of Geography provocatively characterizes this behavior of nature: "everything is related to everything else, but near things are more related than distant things" Tobler (1970, p. 236). This statement is a vivid representation of the phenomenon known as spatial (auto)correlation. Without it, natural phenomena would be disorganized, and physical, geological, and ecological processes, among others, could not take place. On the pragmatic side, if nature did not display spatial structures, there would be no geography or geographers.

Statisticians have long been interested in the description and quantification of spatial structures displayed by data observed in geographic space. These scientists started with a method derived from regression analysis, known as trend surface modeling. Krumbein (1956, 1959) and Grant (1957) first used this method in the earth sciences, following an earlier proposal by "Student" to describe temporal variation using a polynomial function of time ("Student" [Gosset], 1914).

In the early days of spatial analysis by practitioners in various fields, spatial correlation in variables was viewed merely as a nuisance: its presence in data made the usual tests of significance invalid when applied without corrections, whereas modified tests were difficult to implement and had not been fully worked out by statisticians. In my own field of specialty, community ecology, key papers appeared in Levin (1992) and Legendre (1993), arguing that spatial structures were a most important characteristic of the distribution of organisms and natural populations in ecosystems, and were worthy of study for their own sake.

Trend surface analysis, which uses a polynomial function of the geographic coordinates of study sites, was used in early attempts at spatial modeling. This is a rather crude method: to model fine spatial structures would require a polynomial equation with more monomials than observations, which in turn would render the method useless in practice in regression. Researchers then started looking for an applicable method that would produce fine-resolution spatial models with a reasonable number of parameters, one that could be applied to irregularly spaced study sites and be used to model univariate or multivariate response data.

Moran eigenvector spatial filtering: Multiple origins and convergence

A method to model multiscale spatial patterns based on spatial eigenvectors is known as Moran eigenvector spatial filtering (MESF). This book by Griffith, Chun, and Li is about that methodology. Interestingly, this method was developed independently and nearly simultaneously in two different fields, statistical geography (Griffith, 1996, 2000) and quantitative community ecology (Borcard & Legendre, 2002[1]). Following their first paper, Borcard, Legendre, Avois-Jacquet, and Tuomisto (2004) published a series of real-world ecological applications of this method. A few years later, Dray, Legendre, and Peres-Neto (2006) formalized the theory of Moran's eigenvector maps (MEM).

Griffith's original goal was to filter the effect of spatial autocorrelation out of model residuals, transferring this component to a model's conditional mean (i.e., intercept), whereas that of Legendre and his coauthors was to explicitly model the multiscale nature of univariate or multivariate response data[2]. The MESF method of analysis was based on earlier developments by geographers to analyze binary spatial connection (i.e., spatial weights) matrices (SWM; Garrison & Marble, 1964; Gould, 1967; Tinkler, 1972; Griffith, 1996). The two groups quickly realized that their methods had the same algebraic bases and that their objectives were interchangeable. Researchers from these two groups jointly published a paper unifying the terminology and defining the field of *spatial eigenfunction analysis* (Griffith & Peres-Neto, 2006), which encompasses all methods based on eigenvectors describing the spatial relationships among study sites.

Subsequently, Griffith and Legendre had an opportunity to exchange notes in August 2007 during a conference organized by Academia Sinica in Taipei, where they had been invited separately and independently to present their methods. They explained to the audience that the two methods, although formally presented in different ways, were actually one and the same.

[1] I had presented this method two years earlier in a keynote address delivered at the Modelling Complex Systems conference in Montréal in July 2000.

[2] In the fields of community ecology and biogeography, and contrary to many problems in geography, most response datasets are (highly) multivariate and nonnormal.

A word about the theoretical background for MESF in ecology

By asking me to write this "Foreword" for their book, Griffith, Chun, and Li offered me an opportunity to explain the theoretical bases that make ecologists interested in spatial correlation and its modeling by spatial eigenfunctions. For that, I have to go back a little in the history of community ecology. This is the branch of ecology devoted to the scientific study of relationships among the species forming natural communities, as well as relationships between these species and their environmental conditions. In the 1990s, ecologists became aware that different kinds of generating processes could produce spatial correlation in data. The main mechanisms are the following: (1) *induced spatial dependence*: the functional dependence of given response data (e.g., species) on a set of explanatory variables; this process is in action when species forming natural communities are dependent upon the environmental conditions in which they are found; (2) *true autocorrelation*: spatial correlation that may occur in multivariate data because of functional interactions among the species in a multivariate data matrix; and (3) *historical dynamics:* manifestations of past natural events, such as isolation by geographic barriers and disturbances of various kinds (e.g., storms, forest fires, volcanic eruptions, and landslides), and anthropogenic causes, such as agriculture, logging, mining, and constructions of various sizes; these past processes may have caused spatial structures to emerge and may have left traces in present-day data that can be identified and modeled as spatial structures. Researchers in other fields could apply these hypotheses, conceptualizations, or theory elements to the explanation of spatial structures they find in their data.

The methodological developments of MESF to date by quantitative geographers are described in detail in the 10 chapters of this book. Meanwhile, methodological developments of MESF continue in ecological research. Blanchet and his coauthors developed asymmetric spatial eigenvector maps in the late 2000s (Blanchet, Legendre, & Borcard, 2008; Blanchet, Legendre, Maranger, Monti, & Pepin, 2011), a method designed to model the effects of directional physical processes, such as marine and river currents, on ecological communities. Guénard, Legendre, Boisclair, and Bilodeau (2010) decomposed the correlation between variables into spatial scales and then extended their method to multivariate response data matrices (Guénard & Legendre, 2018). Following another research path, Guénard, Legendre, and Peres-Neto (2013) extended the spatial eigenvector framework to the modeling of phylogenetic trees and used MESF eigenvectors to predict different types of traits and properties unobserved in rare or endangered species (Guénard, Boisclair, & Legendre, 2015; Guénard, von der Ohe, Walker, Lek, and Legendre (2014). Spatial eigenvectors also were used to develop a test for space–time interaction in repeated surveys (through time) of sets of sites without replication (Legendre, De Cáceres, & Borcard, 2010).

Simultaneous development of a methodology by researchers in different disciplines is an indication of its strength. The MESF method was developed independently by two groups of researchers, at about the same time, which may provide users of the method more confidence in it. Reviewing the variety of ways MESF analysis has been applied to real-world data by geographers, on the one hand, and by ecologists, on the other hand, may give users in different fields ideas for applications that they initially had not considered.

In addition, ecologists are interested in software and, in particular, in R. They developed an R package devoted to spatial and temporal analysis called adespatial (Dray et al., 2016, the first version released on CRAN), with a strong emphasis on spatial eigenfunction analysis. This software and that presented in especially Chapter 4 of this book provide a powerful toolbox for practitioners.

Scientists who apply MESF methods to analyze data recognize that spatial eigenfunctions, which describe unmeasured spatial relationships among the sites constituting study units, may be a proxy for unmeasured explanatory variables. This perspective means, in practice, that one can use these eigenfunctions to model and predict spatial structures without detailed quantitative knowledge of all explanatory variables affecting a set of observed data.

Extensions and the future of MESF analysis

Extensions of MESF to the analysis of temporal data (time series) and to space–time data appeared in the 2010s. In the Preface of this book, Griffith, Chun, and Li mention these developments, which were the result of the work of several groups of researchers in statistical geography. Ecologists also extended MESF to the analysis of space–time data (Legendre & Gauthier, 2014).

I would like to express my highest appreciation to Professors Griffith, Chun, and Li for the immense amount of work they have completed to produce the 10 chapters of this book. Developers and users of spatial eigenfunction methods in all fields of the natural and social sciences will read, study, and refer to this book, which constitutes the first comprehensive compendium of spatial eigenfunction research. This book represents a necessary and effective step toward future development and diffusion of the method. Furthermore, seeing what new understanding will be achieved by researchers about the role of spatial structures in natural and man-made systems—researchers who now have a firm basis of knowledge upon which to base their applications of MESF and its future developments—will be exciting.

Pierre Legendre
Université de Montréal, Montreal, QC, Canada
April, 2019

References

"Student" [W. S. Gosset]. (1914). The elimination and spurious correlation due to position in time or space. *Biometrika, 10*, 179–180.

Blanchet, F. G., Legendre, P., & Borcard, D. (2008). Modelling directional spatial processes in ecological data. *Ecological Modelling, 215*, 325–336.

Blanchet, F. G., Legendre, P., Maranger, R., Monti, D., & Pepin, P. (2011). Modelling the effect of directional spatial ecological processes at different scales. *Oecologia, 166*, 357–368.

Borcard, D., & Legendre, P. (2002). All-scale spatial analysis of ecological data by means of principal coordinates of neighbour matrices. *Ecological Modelling, 153*, 51–68.

Borcard, D., Legendre, P., Avois-Jacquet, C., & Tuomisto, H. (2004). Dissecting the spatial structure of ecological data at multiple scales. *Ecology, 85*, 1826–1832.

Dray, S., Blanchet, G., Guénard, G., Jombart, T., Legendre, P., & Wagner, H. H. (2016). *adespatial: multivariate multiscale spatial analysis. R package version 0.0-2. http://cran.r-project.org/package=adespatial.*

Dray, S., Legendre, P., & Peres-Neto, P. R. (2006). Spatial modelling: a comprehensive framework for principal coordinate analysis of neighbour matrices (PCNM). *Ecological Modelling, 196*, 483–493.

Garrison, W. L., & Marble, D. F. (1964). Factor analytic study of the connectivity of a transportation matrix. *In vol. 12. Papers and Proceedings, Regional Science Association*, (pp. 231–238).

Gould, P. R. (1967). On the geographical interpretation of eigenvalues. *Transactions of the Institute of British Geographers*, (42), 53–92.

Grant, F. A. (1957). A problem in the analysis of geophysical data. *Geophysics, 22*, 309–344.

Griffith, D. (1996). Spatial autocorrelation and eigenfunctions of the geographic weights matrix accompanying geo-referenced data. *Canadian Geographer, 40*, 351–367.

Griffith, D. (2000). A linear regression solution to the spatial autocorrelation problem. *Journal of Geographical Systems, 2*, 141–156.

Griffith, D., & Peres-Neto, P. R. (2006). Spatial modeling in ecology: the flexibility of eigenfunction spatial analyses. *Ecology, 87*, 2603–2613.

Guénard, G., Boisclair, D., & Legendre, P. (2015). Phylogenetics to help predict active metabolism. *Ecosphere, 6*, art62.

Guénard, G., & Legendre, P. (2018). Bringing multivariate support to multiscale codependence analysis: assessing the drivers of community structure across spatial scales. *Methods in Ecology and Evolution*, *9*, 292–304.

Guénard, G., Legendre, P., Boisclair, D., & Bilodeau, M. (2010). Multiscale codependence analysis: an integrated approach to analyze relationships across scales. *Ecology*, *91*, 2952–2964.

Guénard, G., Legendre, P., & Peres-Neto, P. R. (2013). Phylogenetic eigenvector maps: a framework to model and predict species traits. *Methods in Ecology and Evolution*, *4*, 1120–1131.

Guénard, G., von der Ohe, P. C., Walker, S. C., Lek, S., & Legendre, P. (2014). Using phylogenetic information and chemical properties to predict species tolerances to pesticides. *Proceedings of the Royal Society B*, *281*, 20133239.

Krumbein, W. C. (1956). Regional and local components in facies maps. *Bulletin of the American Association of Petroleum Geologists*, *40*, 2163–2194.

Krumbein, W. C. (1959). Trend-surface analysis of contour type maps with irregular control-point spacing. *Journal of Geophysical Research*, *64*, 823–834.

Legendre, P. (1993). Spatial autocorrelation: trouble or new paradigm? *Ecology*, *74*, 1659–1673.

Legendre, P., De Cáceres, M., & Borcard, D. (2010). Community surveys through space and time: testing the space–time interaction in the absence of replication. *Ecology*, *91*, 262–272.

Legendre, P., & Gauthier, O. (2014). Statistical methods for temporal and space–time analysis of community composition data. *Proceedings of the Royal Society B*, *281*, 20132728.

Levin, S. A. (1992). The problem of pattern and scale in ecology. *Ecology*, *73*, 1943–1967.

Tinkler, K. J. (1972). The physical interpretation of eigenfunctions of dichotomous matrices. *Transactions of the Institute of British Geographers*, (55), 17–46.

Tobler, W. (1970). A computer movie simulating urban growth in the Detroit Region. *Economic Geography*, *46*, 234–240. Supplement.

Preface

Moran eigenvector spatial filtering (MESF) is a relatively recent novel and powerful statistical methodology that accounts for spatial autocorrelation (SA) in its georeferenced data analyses. Its appeal is its simplicity in terms of regression analysis. Its implementation drawbacks include serious complexities associated with constructing an eigenvector spatial filter (ESF). The principal goal of this book is to provide an accessible reference book for applying MESF to spatial regression modeling by including ESFs in spatial regression model specifications. It interfaces with user-friendly software primarily developed by Dr. Hyeongmo Koo. The purpose of this preamble is to furnish an overview of the history of and motivation for developing MESF, as well as an overview of the georeferenced Texas dataset used for empirical illustrations throughout this book.

Setting the stage for MESF, Cliff and Ord (1973) published their book entitled *Spatial Autocorrelation*, initiating a popularization of this fundamental concept as well as the spatial auto-normal probability model in geographic information science (GIScience) and the geographic/geospatial sciences. Besag (1974) published "Spatial interaction and the statistical analysis of lattice systems," extending the spatial auto-models beyond the auto-normal specification. These two publications highlighted important emerging spatial statistical methodology challenges: (1) the intractability of the auto-normal model log-Jacobian term (i.e., the normalizing constant); (2) the difficulty of specifying auto-models for nonnormal data (which eventually were implemented with the numerically intensive Markov chain Monte Carlo [MCMC] technique); and (3) the failure of certain auto-models to account for positive SA (PSA; e.g., the auto-Poisson and the auto-negative binomial). Meanwhile, considerable work in the late 1970s demonstrated that SA mattered and contrasted georeferenced data analyses ignoring SA with those acknowledging SA.

This latter theme became the topic of Griffith's doctoral dissertation, in which he established the rudimentary foundation of MESF (Griffith, 1978a). This foundation involved the spatial auto-normal model, in its simultaneous autoregressive (SAR) specification, and filtered SA from georeferenced data in a way that parallels the Cochrane–Orcutt filtering of time-series data containing serial correlation (i.e., temporal autocorrelation). Comparisons of analysis results for georeferenced data based on original and filtered data reveal data analytic complications attributable to SA. Three journal articles derived from this dissertation highlight this approach: a 1978 *Geographical Analysis* paper about ANOVA, a 1979 *Economic Geography* paper involving an empirical data analysis, and a 1981 *Geographical Analysis* rejoinder reporting before and after spatial filtering results for multivariate statistical analyses of georeferenced data (Griffith, 1978b, 1979, 1981). The idea of spatial filtering was formulated at this time but without being directly linked to the notion of a spatial weights matrix (SWM) or the Moran coefficient (MC).

Next, Griffith turned his research attention and efforts to SWMs, applying his knowledge about principal components analysis (PCA) to them. Inspiration for this work mostly derived from exposure to PCA applications to spatial interaction data done by his former University of Toronto professors, James Simmons and Larry Bourne, during and shortly after his doctoral program of study. In his attempt to further articulate spatial filtering, Griffith (1984) published a PCA study of a SWM. His results were disappointing but did emphasize that the principal eigenfunction of a SWM, which already had been promoted in transportation geography as a quantification of topological accessibility associated with a geographic configuration of areal units (based upon the dual graph of its corresponding surface partitioning), offered promise. Accordingly, Griffith (1988) published a chapter in his spatial statistics book showing that the principal eigenvector of a SWM highlights a dimension that spans socioeconomic/demographic, spatial interaction, and land use data. Interestingly, Boots and Kanaroglou (1988) had exactly the same idea about the principal eigenvector being a useful regression analysis covariate, also publishing their paper in that same year.

de Jong, Sprenger, and van Veen (1984) published an article emphasizing selected relationships between a SWM and the MC and the Geary ratio (GR) indices of SA. Griffith (1988) contemplated this and the contents of his work for about a decade. Then, Tiefelsdorf and Boots (1995) published their paper unmasking the relationship between the eigenvalues of a SWM and SA. This was the breakthrough that Griffith needed to progress beyond his 1984 paper. He immediately recognized the connection of their work to the eigenvectors of a SWM and published his seminal and instrumental *Canadian Geographer* paper in 1996. This paper shaped the unfolding of MESF methodology, ultimately resulting in Griffith's 2003 book entitled *Spatial Autocorrelation and Spatial Filtering: Gaining Understanding through Theory and Scientific Visualization*, which furnishes an overview of the theory and practice of MESF.

Data analytic advantages of MESF include that it (1) avoids the need to deal with nonstandard normalizing constants of probability density/mass function (e.g., the log-Jacobian term in an auto-normal model specification); (2) allows spatial generalized linear models to be estimated without the need for MCMC techniques (Griffith, 2002); (3) allows all auto-model contexts to account for PSA; and (4) supports spatially varying coefficients' model specifications (Griffith, 2008). Interestingly, Legendre and his colleagues began formulating a similar spatial statistical methodology several years after the publication of Griffith's 1996 paper (see, e.g., Borcard & Legendre, 2002). Their technique constructs a SWM from the geographic distances among areal units (e.g., the point locations of study sites) and then truncates it at an appropriately chosen threshold that keeps all locations connected. This conceptualization is more reminiscent of geostatistical than spatial autoregressive concepts and works with eigenfunctions extracted from the truncated distance matrix. Recognition of the parallel developments resulted in an *Ecology* paper by Griffith and Peres-Neto (2006) demonstrating that, except for trivial details, these two methodologies are the same. Meanwhile, Tiefelsdorf and Griffith (2007) bolstered the formalization of MESF theory; in the 2000s, Griffith and Chun separately and jointly, along with other researchers, began extending MESF to spatial interaction contexts (e.g., Chun & Griffith, 2011); and, in the early 2010s, spatial scientists undertook major extensions of MESF to space–time contexts (e.g., Patuelli, Schanne, Griffith, & Nijkamp, 2012), and Griffith and Chun (2014, 2016a, 2016b) collaboratively began more fully developing aspects of MESF. With the need for user-friendly implementation in mind, Li initiated a project at the State Key Laboratory of Information Engineering in Surveying, Mapping, and Remote Sensing (LIESMARS) at Wuhan University, China, to provide MESF service through Microsoft's Azure Cloud.

Today, MESF is widely employed across the spatial sciences. For example, Griffith's 2003 book (with 658 citations according to Google Scholar, as of 7/6/2019) is cited in papers published in the following journals from disparate disciplines:

> *American Naturalist, Annals of the AAG, Brazilian Journal of Botany, Demographic Research, Diversity and Distributions, Ecology, Estuaries and Coasts, Genetica, Geomorphology, GIScience and Remote Sensing Health Economics, Informatica, Journal of Applied Econometrics, Journal of Geographical Systems, Journal of Hydrometeorology, Journal of the Royal Statistical Society, Landscape Ecology, Lecture Notes in Computer Science, Papers in Regional Science, Population Research and Policy Review, and Spatial Economic Analysis.*

This list is an illustrative rather than exhaustive enumeration. Furthermore, Chun & Griffith's 2013 book, entitled *Spatial Statistics & Geostatistics*, is an outgrowth of a 2008 US National Science Foundation-funded workshop convened at the University of Texas at Dallas to disseminate MESF methodology. This book's goal is to promote a better understanding of how to implement MESF methodology. Accordingly, it includes comparisons of geostatistics, spatial autoregression, and MESF analyses. It also includes R code for implementing the presented analyses (for chapter-by-chapter digital versions of this code, see http://www.sagepub.com/chun_griffith/study/default.htm). Unfortunately, this code still requires some computer programming skills. Our current book is devoted solely to MESF-based regression analysis and the pedagogy for constructing ESFs, and as such is accompanied by the Spatial Analysis using ArcGIS Engine and R (SAAR) (http://TheSAAR.github.io) and ESF Tool (https://github.com/esftool/esftool; Koo, Chun, & Griffith, 2018) software that provide a user-friendly implementation of MESF.

The future holds considerable promise for MESF because this methodology relates to contemporary work occurring in statistics pertaining to dimension reduction models. In addition, it supports nonnormal data analyses, as well as random effects and both parametric and multiple distributions mixture models. It can be implemented in a frequentist or a Bayesian context. Finally, it supports the scientific visualization of SA.

Data description

Texas county resolution American Community Survey (ACS) data for 2010–2014 furnish the basis for empirical examples throughout this book. These data include 21 attribute variables (Table P1) for the 254 Texas counties (Fig. P1). For spatial interaction analysis illustrations, the datasets also include the five intercounty flows-related attribute variables listed in Table P2. These data and selected computer code are available through https://github.com/ywchun/sp_esf or http://bit.ly/sp_esf.

According to these data, the total population of Texas for this time period is 26,092,033, with county populations ranging from 89 to 4,269,608. This range implies that these data potentially present some small number issues for certain types of data analyses. These data also reveal that Texas has a near-equal number of men and women, with 49.6% of the population classified as male and 50.4% classified as female. The predominant racial composition of its population is as follows: 74.7% Caucasian, 11.9% African American, and 4.1% Asian (9.3% other). Among the ethnic groups,

Table P1 Selected Texas county resolution attribute variables for 2010–14

ACS field	Acronym	Description
Pop_Total	TP	Total population
Pop_Male	TPM	Male population
Pop_Female	TPF	Female population
Pop_White	TPW	People who are white only
Pop_Black	TPB	People who are black or African American only
Pop_Asian	TPA	People who are Asian only
Pop_Hisp	TPH	Total Hispanic or Latino population
Com_Total16Over	WKR	Total workers 16 years and older
Com_SumTravelMin	TT	Aggregate travel time to work (in minutes) of workers
HH_Total	TH	Total households
Edu_25Older	P25	Population 25 years and older
Edu_HSD_Higher	PED	Population with a high school diploma or higher education
Pov_PopEval	TPPV	Total population for whom poverty status is determined
Poverty	PPV	Population whose income in the past 12 months was below the poverty level
Inc_MedHH	MINC	Median household income in the past 12 months (in 2014 inflation-adjusted dollars)
FoodStamps	FS	Households receiving food stamps/SNAP in the past 12 months
Uemp_16Older	TUNE	Total population 16 years and older
Uemp_NotLabor	UNEMP	Population 16 years and older not in the labor force
Housing_Total	HT	Total housing units
House_OwnerOcc	HOO	Owner occupied housing units
House_Vacant	HV	Vacant housing units

Hispanic dominates and accounts for 38.2% of the population. The median annual household income by county ranges from $22,176 to $86,597, with 17.7% of the population classified as being in poverty. Of the people at least 16 years of age, 58.8% participate in the state's labor force. In addition, of the people at least 25 years of age, 85.6% have a high school diploma or post-high school education. Finally, 55.5% of the statewide housing stock is owner occupied and 11.5% of it is vacant.

For simplicity, intercounty Euclidean distances were computed with latitude and longitude coordinates of the county centroids (see Fig. P1); these

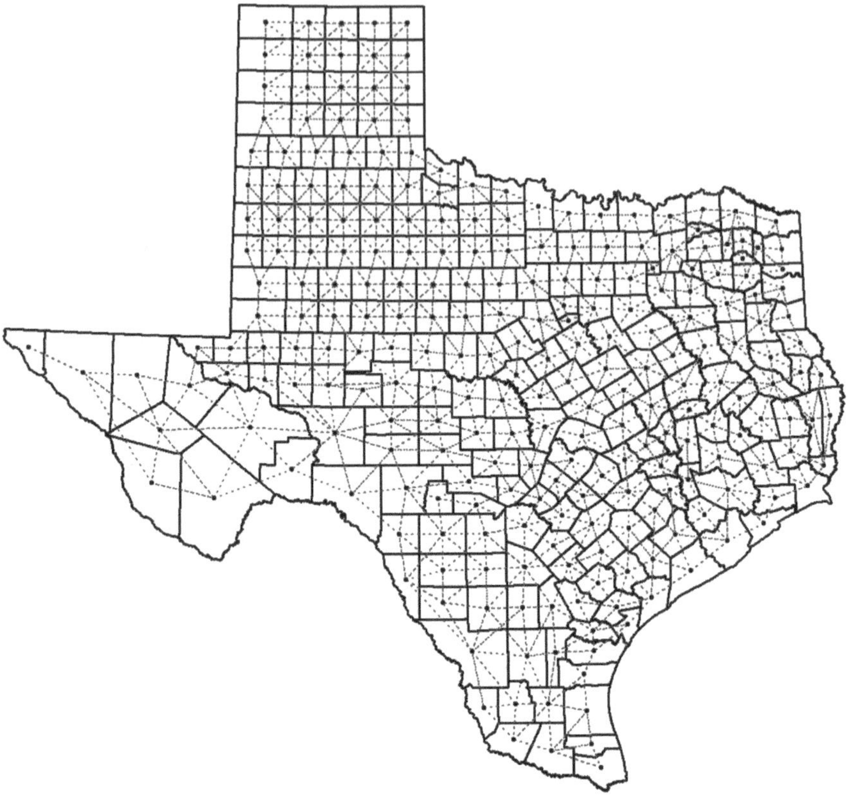

Fig. P1 The 254 counties of Texas during 2010–14, together with their geometric centroids (*black points*) and queen's adjacency topological structure dual graph (*dotted lines*).

Table P2 Spatial interaction intercounty 2010–14 migration flows- and 2009–13 journey-to-work flows-related attribute variables

ACS field	Acronym	Description
FPopulation	FRP	Population at an origin
Fnonmover	FNM	People at an origin who did not move in the last 12 months
TPopulation	TOP	Population at a destination
Tnonmover	TNM	People at a destination who did not move in the last 12 months
Inter_county_movers	FLOWS	Number of migrants
Residence County	FCFIPS	Origin county
Place of Work County	TCFIPS	Destination county
Workers in Commuting Flow	T_ij	Journey-to-work flow

distances deviate negligibly from their great-circle distance counterparts. In this Texas context, the maximum discrepancy between Euclidean and great-circle (i.e., orthodromic) distance is for the separation measurement between Cameron and Dallam Counties (i.e., the northwestern corner of the panhandle and the southeastern tip adjacent to the Mexican border and the Gulf of Mexico).

A preview of the book's content

The preceding section describes the Texas dataset used throughout this book to illustrate various regression analyses employing MESF conceptualizations and implementations. This book contains 10 chapters that address the specifics of these MESF-based regression analysis features.

Chapter 1, entitled "Spatial autocorrelation," furnishes the motivation for engaging in and the background to MESF. It begins by defining the concept of SA, couching it within the more general context of correlated data, and then extending this context to embrace Tobler's (1970) First Law of Geography. In classical statistics, correlated data refer to such situations as repeated measures (which essentially introduce a time-series component to data) and paired observations (e.g., biological twins); the organizational structure of these data is sequencing, for time-series types of data, and order-not-important similarity for paired data. This extension to geographic data is for correlation arising because observations occupy nearby locations; n 0–1 indicator variables, one for each of n locations, denoting whether two locations are neighbors, define the organizational structure of a SWM. As with conventional statistical analysis, gauging SA relates to the measurement scale (i.e., nominal, ordinal, interval/ratio) used to quantify georeferenced random variables (RVs) as well as to the idea of similarity used to quantify correlation. MESF is a novel and powerful spatial statistical methodology that allows spatial scientists to account for SA in their georeferenced data analyses. Overall, the MC, which is the basis of MESF, tends to be the best of the popular SA indices. Accordingly, a section of this chapter summarizes its statistical distribution theory. An ensuing section reviews impacts of SA on RV frequency distributions, highlighting that its principal impact is on variance. This section also demonstrates complications attributable to deviations from independent observations and the usefulness of SA visualization with such tools as the Moran scatterplot. This chapter concludes with a stand-alone appendix that presents mathematical details about the MC for linear regression residuals.

Chapter 2, entitled "An introduction to spectral analysis," lays the mathematical and technical foundation for MESF, which involves the spectral decomposition of SWMs. Relevant quantities in a SWM spectrum—which relate to specific natures and degrees of, as well as map patterns accompanying, particular manifestations of SA—are eigenfunctions. Each eigenfunction is a pair of mathematical quantities, a scalar known as an eigenvalue, and an n-by-1 vector, one element for each location, known as an eigenvector. Most quantitative spatial scientists first encounter eigenfunctions when studying multivariate statistics, commonly with PCA, a conceptualization that furnishes a useful perspective for introducing this subject. Next, this chapter summarizes spectral analysis applications involving one-, two-, and three-dimensional data, respectively, exploiting time series, spatial series, and space–time series analyses to better explain a SWM spectrum. This latter discussion is a building block for the content of Chapter 8.

Chapter 3, entitled "MESF and linear regression," presents the formulation of MESF for linear regression and Gaussian (i.e., normal) RVs. This context was the first to be developed following publication of Griffith's seminal 1996 paper establishing MESF because linear regression is the workhorse of classical statistics, and normal curve theory furnished the initial context for spatial autoregression developments. While reviewing the formulation of MESF, this chapter summarizes comparisons between ESF and the SAR model specification. It also outlines a theoretical mathematical statistical foundation for ESF construction and posits the fundamental theorem of MESF. The emphasized interpretation of SA is map pattern, with connections made to heterogeneity in map-wide trends. The MESF linear model specification enables estimation of an ESF with the ordinary least squares (OLS) technique. It also allows articulation of relationships between the standard suite of model diagnostics and a constructed ESF. A selection of eigenvectors for ESF construction involves stepwise procedures (Chun, Griffith, Lee, & Sinha, 2016), which are legitimate because the eigenvectors in question are extracted from a modified SWM, with this modification resulting in mutually orthogonal and uncorrelated eigenvectors. This chapter concludes with a discussion about interpreting a constructed ESF as well as its regression coefficients that determine its linear combination of eigenvectors, and a discussion about ESF prediction in terms of linear regression. One advantage of MESF noted here concerns its ability to bolster simulation experiments containing georeferenced data.

Chapter 4, entitled "Software implementation for constructing an ESF, with special reference to linear regression," addresses serious complexities associated with implementing the construction of an ESF; the most difficult task confronting spatial scientists in practice appears to be the extraction of the correct eigenfunctions. This chapter seeks to furnish a practical guide for implementing MESF with user-friendly software developed by the authors to support and simplify its implementation through the SAAR and ESF Tool software. The computer packages discussed in this chapter, coupled with the sample datasets presented in the preceding section, allow replication of illustrative analyses found throughout this book and furnishes a template for other spatial scientists to use for their own data analyses. This chapter offers additional guidance about geographic scale and resolution issues associated with ESFs, determination of the candidate set of eigenvectors for constructing an ESF, and extensions of MESF to large georeferenced datasets. Notably, two ESF modules already exist in the R project (i.e., the SpatialFiltering and ME functions in the *spdep* package and the *spmoran* package).

Chapter 5, entitled "MESF and generalized linear regression," expands the range of RVs discussed in the Gaussian RV section of Chapter 3. The most common RVs dealt with by spatial scientists are the binomial, with its Bernoulli special case of presence/absence (i.e., logistic), and the Poisson, with its overdispersion case captured by the negative binomial. These nonnormal RVs treated in this chapter require nonlinear regression techniques to estimate their parameters. SA further complicates these data analyses, requiring intractable normalizing constants (i.e., probability density/mass function constants ensuring that all possible probabilities for a given RV integrate/sum to 1). Paralleling the Bayesian approach, which also is plagued by the normalizing constant complication, parameter estimation needs to be done with MCMC sampling techniques, which are extremely numerically intensive. MESF avoids this necessity. Furthermore, autoregressive model specifications for the Poisson and negative binomial, among others, are unable to describe landscapes in which PSA prevails. An appropriately constructed ESF is able to do this. This chapter concludes with a description of ESF prediction using generalized linear models regression.

Chapter 6, entitled "Modeling spatial heterogeneity with MESF," goes beyond the more familiar spatial statistical conceptualization of SA as a trend to the more familiar spatial econometrics conceptualization of SA as heterogeneity across a geographic landscape. This subtle shift in viewpoint alludes to geographically varying coefficients in spatial regression models. One prevalent specification, other than the standard inclusion of covariates, is to capture this geographic heterogeneity with a spatially varying intercept term; this is exactly what the spatial lag term does in a spatial autoregressive model specification. This also is what an ESF does: it filters out SA in residuals and transfers it to the intercept term. Covariate coefficients also can have geographically varying coefficients, which the SAR model specification has. Meanwhile, ESFs also can be constructed for covariate regression coefficients. One criticism about geographically weighted regression (GWR), which accentuates geographically varying coefficients, is the presence of multicollinearity in these nonconstant coefficients (Wheeler and Tiefelsdorf, 2005). The eigenvectors constituting coefficient ESFs clarify this situation: regression coefficients with common eigenvectors tend to inflate pairwise correlation, whereas regression coefficients with unique eigenvectors tend to deflate pairwise correlation. This chapter also includes a discussion of conventional statistical random effects terms, which often are geographically varying intercept terms with a spatially unstructured component that is aspatial and a spatially structured component that relates to latent SA. Chapter 6 concludes with a section entitled "Local SA ESFs" that extends the notion of heterogeneity examined in the first part of the chapter. Here the heterogeneity concerns global measures of SA, which may have anomalous local deviations from their global counterparts. The two most popular SA indices—the MC and the GR—have local versions. In addition, the Getis–Ord statistic, which more closely aligns with geostatistics (although the GR relates to geostatistics, too), also has a popular local version. This chapter furnishes descriptions of empirical relationships between these three indices and ESFs. As such, its summarized evaluation is more exploratory in nature.

Chapter 7, entitled "Spatial interaction modeling," shifts attention from more static cross-sectional datasets to more dynamic flows datasets. An emphasis on the important role SA plays in spatial interaction modeling dates as far back as Curry (1972). Successfully accounting for SA in spatial interaction model specifications is a recent achievement of the spatial sciences, dating back to around 2007. This chapter uses migration and journey-to-work movements to illustrate how ESFs account for SA in flows data.

Chapter 8, entitled "Space–time modeling," moves even further from static datasets to space–time datasets. It broadens the discussion to MESTF (i.e., Moran eigenvector space–time filtering), where T denotes time. MESTF integrates a time-series structure with a space-series structure (i.e., a SWM). Two possibilities exist: the space–time lag structure specification and the space–time contemporaneous structure specification. Paralleling Chapters 3 and 5, this chapter presents results for ESTF prediction.

Chapter 9, entitled "MESF and multivariate statistical analysis," focuses on both common and unique map pattern sources of SA spanning multiple attribute variables. The former tend to inflate product moment, Spearman's rank, and other coefficients of correlation calculated for two geographic distributions of attribute values, whereas the latter tend to shrink such coefficients toward zero. One outcome is impacts on multicollinearity that exacerbate or suppress it, with consequences about orthogonality for PCA, factor analysis, and canonical correlation analysis. In addition to multicollinearity complications, PSA induces a tendency for similar attribute values to concentrate regionally in a geographic landscape, reducing within regions variances, and increasing between regions variances, impacting multivariate analysis of variance and discriminant function analysis evaluations. Especially this map pattern feature of SA can have a profound influence on cluster analysis results. Moreover, this chapter demonstrates that SA matters in multivariate data analysis, and that MESF furnishes a tool for better understanding how SA matters in this context.

Finally, Chapter 10, entitled "Concluding comments: toy dataset implementation demonstrations," delivers computer screenshots for a number of illustrative implementations of MESF using the SAAR software. The toy dataset it uses comprises 12 Texas counties that formed the Dallas–Fort Worth metropolitan region in 2010. The accompanying attribute table includes population, area, Hispanic, and total population counts, and county centroid geocodes (i.e., latitude and longitude). This chapter allows readers to verify their understanding about how to execute MESF regression.

Daniel A. Griffith
Richardson, TX
Yongwan Chun
Richardson, TX
Bin Li
Mt. Pleasant, MI
July, 2019

References

Besag, J. (1974). Spatial interaction and the statistical analysis of lattice systems. *Journal of the Royal Statistical Society. Series B, Methodological*, *36*(2), 192–236.

Boots, B., & Kanaroglou, P. (1988). Incorporating the effects of spatial structure in discrete choice models of migration. *Journal of Regional Science*, *28*, 495–509.

Borcard, D., & Legendre, P. (2002). All-scale spatial analysis of ecological data by means of principal coordinates of neighbour matrices. *Ecological Modelling*, *153*, 51–68.

Chun, Y., & Griffith, D. (2011). Modeling network autocorrelation in space–time migration flow data: An eigenvector spatial filtering approach. *Annals of the AAG*, *101*, 523–536.

Chun, Y., & Griffith, D. (2013). *Spatial statistics and geostatistics*. Thousand Oaks, CA: Sage.

Chun, Y., Griffith, D., Lee, M., & Sinha, P. (2016). Eigenvector selection with stepwise regression techniques to construct eigenvector spatial filters. *Journal of Geographical Systems, 18*(1), 67–85.
Cliff, A. D., & Ord, J. K. (1973). *Spatial autocorrelation*. London: Pion.
Curry, L. (1972). A spatial analysis of gravity flows. *Regional Studies, 6*, 131–147.
de Jong, P., Sprenger, C., & van Veen, F. (1984). On extreme values of Moran's I and Geary's c. *Geographical Analysis, 16*, 17–24.
Griffith, D. (1978a). *The impact of configuration and spatial autocorrelation on the specification and interpretation of geographical models*. PhD dissertation Department of Geography, University of Toronto.
Griffith, D. (1978b). A spatially adjusted ANOVA model. *Geographical Analysis, 10*, 296–301.
Griffith, D. (1979). Urban dominance, spatial structure and spatial dynamics: Some theoretical conjectures and empirical implications. *Economic Geography, 55*, 95–113.
Griffith, D. (1981). Towards a theory of spatial statistics: A rejoinder. *Geographical Analysis, 13*, 91–93.
Griffith, D. (1984). Measuring the arrangement property of a system of areal units generated by partitioning a planar surface. In G. Bahrenberg, M. Fischer, & P. Nijkamp (Eds.), *Recent Developments in Spatial Analysis: Methodology, Measurement, Models* (pp. 191–200). Aldershot: Gower.
Griffith, D. (1988). *Advanced spatial statistics*. Dordrecht: Martinus Nijhoff.
Griffith, D. (1996). Spatial autocorrelation and eigenfunctions of the geographic weights matrix accompanying geo-referenced data. *The Canadian Geographer, 40*, 351–367.
Griffith, D. (2002). A spatial filtering specification for the auto-Poisson model. *Statistics & Probability Letters, 58*, 245–251.
Griffith, D. (2003). *Spatial autocorrelation and spatial filtering: Gaining understanding through theory and scientific visualization*. Berlin: Springer-Verlag.
Griffith, D. (2008). Spatial filtering-based contributions to a critique of geographically weighted regression (GWR). *Environment & Planning A, 40*, 2751–2769.
Griffith, D., & Chun, Y. (2014). Spatial autocorrelation and eigenvector spatial filtering. In M. Fischer & P. Nijkamp (Eds.), *Handbook of regional science* (pp. 1477–1507). Berlin: Springer-Verlag.
Griffith, D., & Chun, Y. (2016a). Evaluating eigenvector spatial filter corrections for omitted georeferenced variables. *Econometrics, 4*, 29.
Griffith, D., & Chun, Y. (2016b). Spatial autocorrelation and uncertainty associated with remotely-sensed data. *Remote Sensing, 8*, 535. https://doi.org/10.3390/rs8070535.
Griffith, D., & Peres-Neto, P. R. (2006). Spatial modeling in ecology: The flexibility of eigenfunction spatial analyses. *Ecology, 87*, 2603–2613.
Koo, H., Chun, Y., & Griffith, D. (2018). Integrating spatial data analysis functionalities in a GIS environment: Spatial analysis using ArcGIS engine and R (SAAR). *Transactions in GIS, 22*, 721–736.
Patuelli, R., Schanne, N., Griffith, D., & Nijkamp, P. (2012). Persistence of regional unemployment: Application of a spatial filtering approach to local labor markets in Germany. *Journal of Regional Science, 52*, 300–323.
Tiefelsdorf, M., & Boots, B. (1995). The exact distribution of Moran's I. *Environment & Planning A, 27*, 985–999.
Tiefelsdorf, M., & Griffith, D. (2007). Semiparametric filtering of spatial autocorrelation: The eigenvector approach. *Environment & Planning A, 39*, 1193–1221.
Tobler, W. (1970). A computer movie simulating urban growth in the Detroit region. *Economic Geography, 46*(Supplement), 234–240.
Wheeler, D., & Tiefelsdorf, M. (2005). Multicollinearity and correlation among local regression coefficients in geographically weighted regression. *Journal of Geographical Systems*, 7(2), 161–187.

CHAPTER 1

Spatial autocorrelation

Chapter outline

An underlying assumption of classical statistics is independent observations, a postulation allowing the multiplication of marginal probability density/mass functions to establish joint distributions, thus simplifying statistical calculations. Early statisticians recognized that this assumption fails to apply in many real-world situations in which correlations exist among observations. One reconceptualization outcome is correlated samples statistical theory, which addresses correlations between paired/matched observations. Another is time-series analysis, which takes into account correlations between attributes for sequential points in time. A third outcome is spatial-series analysis, which deals with correlation between attributes for nearby locations. This latter notion is the topic of this chapter and a primary incentive for spatial scientists to employ the Moran eigenvector spatial filtering (MESF) methodology in their empirical analyses.

Correlated zero-one identically distributed (i.e., $p_1 = p_2 = p$) Bernoulli random variables (BRVs) furnish an informative illustration here, linking correlated samples theory with spatial-series analysis (see Section 1.1.3). Let ρ denote the correlation between two BRVs, which must be nonnegative here (switching the zero-one definition for one of the two BRVs accommodates negative correlation), and p denote the probability of one

Spatial Regression Analysis Using Eigenvector Spatial Filtering
https://doi.org/10.1016/B978-0-12-815043-6.00001-X

occurring in a random draw from either BRV (i.e., a homogeneous process); $1-\mathrm{p}$ denotes the probability of zero occurring. Correlation between two such BRVs means that if zero is selected in a random draw for the first random variable (RV) with probability $1-\mathrm{p}$, then one occurs in a random draw from the second BRV with probability $(1-\rho)\mathrm{p}$. Meanwhile, if one is selected in a random draw for the first RV with probability p, then one occurs in a random draw from the second BRV with probability $(1-\rho)\mathrm{p}+\rho$. Consequently, the accompanying contingency table has the following format (after Teugels, 1990):

BRV	0	1	Marginal probability
0	$(1-\mathrm{p})[1-(1-\rho)\mathrm{p}]$	$(1-\rho)\mathrm{p}(1-\mathrm{p})$	$1-\mathrm{p}$
1	$(1-\rho)\mathrm{p}(1-\mathrm{p})$	$\mathrm{p}[(1-\rho)\mathrm{p}+\rho]$	p
Marginal probability	$1-\mathrm{p}$	p	1

If $\rho=0$, the cross-tabulation cells in this contingency table contain the standard independence expressions of $(1-\mathrm{p})^2$, $\mathrm{p}(1-\mathrm{p})$, and p^2 familiar to researchers who utilize chi-square analysis. If $\rho=1$, these cross-tabulation cells contain $(1-\mathrm{p})$, zero, and p (i.e., an initial random draw of zero means the second random draw always is zero, and an initial random draw of one means the second random draw always is one). This BRV example exemplifies how correlation alters data and is the basis for the analysis of presence/absence maps (see Fingleton, 1986; Rogerson, 1998). It also furnishes insights into the multiple testing problem.

Accounting for spatial autocorrelation (SA) when it is the source of ρ in the preceding contingency table cross-tabulation requires appropriate modifications to associated probability model specifications. Cliff and Ord (1973), followed by Besag (1974), promoted the use of autoregressive model formulations. These particular specifications have serious weaknesses, including mathematical intractabilities, a need for numerically intensive calculations, and restrictions on the nature of SA accommodated by some of them. Such serious drawbacks constituted a primary impetus for the development of an alternative spatial statistical methodology for handling complications arising from SA latent in georeferenced data. MESF has emerged as one such alternative methodology. Motivation for using MESF to undertake georeferenced data analyses is multifold. Foremost is its simplicity: it works with the entire battery of standard statistical techniques. With regard to the spatial autoregression limitations, it circumvents mathematical intractabilities (e.g., it does not require new normalizing constants for reformulated probability density/mass functions), it reduces the numerical intensity of certain spatial statistical calculations [e.g., it does not require the calculation of eigenvalues or the utilization of Markov chain Monte Carlo (MCMC) techniques], and it can accommodate positive (e.g., for autocorrelated Poisson, negative binomial, and exponential RVs), negative, or both positive and negative SA (NSA) mixtures latent in georeferenced data. In addition, MESF allows visualization of SA for individual georeferenced datasets. Meanwhile, a constructed eigenvector spatial filter (ESF) estimator displays many, if not most or all, of the properties established for spatial autoregression estimators: unbiasedness, efficiency, consistency, correcting for omitted variable bias, and missing data imputation, among others. Furthermore, MESF is flexible, more so than spatial autoregression. Not only can MESF be used in either frequentist or Bayesian types of analyses, but it also can account for SA in spatial flows, and it relates to spatially varying coefficient, random effects, and parametric mixture probability model specifications. Overall, its many appealing features should persuade spatial scientists to adopt MESF; one goal of this book is to furnish them with the particulars to justify such an adoption. The first of these specifics is a thorough treatment of the SA concept, the focus of this chapter.

1.1 Defining SA

The presence of nonzero correlation results in one RV, Y, in a pair being dependent on the other RV, X. Its global trend depicting a positive relationship is for larger values of X and Y to tend to coincide, for intermediate values of X and Y to tend to coincide, and for smaller values of X and Y to coincide; values of X are directly proportional to their corresponding values of Y. For an indirect (i.e., negative or inverse) relationship, larger values of X tend to coincide with smaller values of Y, intermediate values of X and Y tend to coincide, and smaller values of X tend to coincide with larger values of Y; values of X are inversely proportional to their corresponding values of Y. A random relationship has values of X and Y haphazardly coinciding according to their relative magnitudes.

Autocorrelation transfers this notion of relationships from two RVs to a single RV; the prefix auto means self. Accordingly, n observations have $n(n-1)$ possible pairings, one between each observation and the $(n-1)$

remaining observations; each observation always has a correlation of one with itself, and hence these n self-pairings are of little or no interest in terms of correlation. One relevant question asks whether or not an ordering exists that differentiates between two subsets of these $n(n-1)$ pairings such that the ordered subset contains directly correlated observations, whereas the unordered subset contains uncorrelated observations. For data linked to a map, its spatial ordering of attribute values virtually always yields a collection of correlated observations. Because the ordering involved is spatial, the descriptive phrase applied to this situation is SA.

SA may be defined generically as the arrangement of attribute values on a map for some RV Y such that a map pattern becomes conspicuous by visual inspection. More specifically, positive SA (PSA)—overwhelmingly the most commonly observed type of SA—may be defined as the tendency for similar Y values to cluster on a map. In other words, larger values of Y tend to be surrounded by larger values of Y, intermediate values of Y tend to be surrounded by intermediate values of Y, and smaller values of Y tend to be surrounded by smaller values of Y. In contrast, NSA—a rarely observed type of SA—may be defined as the tendency for dissimilar Y values to cluster on a map. In other words, larger values of Y tend to be surrounded by smaller values of Y, intermediate values of Y tend to be surrounded by intermediate values of Y, and smaller values of Y tend to be surrounded by larger values of Y. The absence of SA indicates a lack of map pattern and a haphazard mixture of attribute values across a map.

A number of authors present explanatory definitions of SA, including Getis (2008), Goodchild (1986), Griffith (1987, 1992, 2005, 2009, 2016), Legendre (1993), and Odland (1988).

1.1.1 A mathematical formularization of the first law of geography

The preceding definition of SA indicates that this concept exists because orderliness, (map) pattern, and systematic concentration, rather than randomness, epitomize real-world geospatial phenomena. Tobler's (1969, p. 7) First Law of Geography captures this notion: "everything is related to everything else, but near things are more related than distant things." In 2004 the *Annals of the American Association of Geographers* published commentaries by six prominent geographers (Sui, Barnes, Miller, Phillips, Smith, and Goodchild), together with a reply by Tobler (vol. 94: pp. 269–310) about this notion. Subsequent quantitative SA measurements (e.g., see Section 1.1.3) are mathematical abstractions of this empirical rule.

1.1.2 Quantifying spatial relationships: The spatial weights matrix

A spatial weights matrix (SWM) is an n-by-n nonnegative (i.e., all of its entries are zero or positive) matrix, say **C**, describing the geographic relationship structure latent in a georeferenced dataset containing n observations (areal units or point locations in the case of georeferenced data), and has $n(n-1)/2$ potential pairwise, symmetric relationship designations; without invoking symmetry, it has $n(n-1)$ potential relationship designations. Classical statistics assumes that these pairwise relationship designations do not exist (i.e., observation independence). Time-series analyses assume that $(n-1)$ of these pairwise relationship designations are nonzero and asymmetric (dependence is one-directional in time), with perhaps several additional relationship designations to capture seasonality effects. Spatial data mostly assume that between $n-1$ and $3(n-2)$ of these pairwise relationship designations are nonzero and symmetric, with asymmetric relationships usually specified from symmetric ones. The relationship definition rule (often called the neighbor or adjacency rule) is that correlation between attribute values exists for areal unit polygons sharing a common nonzero length boundary (i.e., the rook definition, using a chess move analogy). One extension of this definition is to nonzero length (i.e., point contacts) shared boundaries (i.e., the queen definition, using a chess move analogy). This latter extension tends to increase the number of designated pairwise correlations for administrative polygon surface partitionings by roughly 10%; its asymptotic upper bound is a doubling of pairwise relationships (i.e., the regular square lattice case) for this near-planar situation, which still constitutes a very small percentage of the $n(n-1)/2$ possible relationships. A third extension is to $k > 1$ nearest neighbors, which fails to guarantee a connected dual graph structure and for which k is sufficiently small that it still constitutes a very small percentage of the $n(n-1)$ possible relationships. In all of these specifications of matrix **C**, if areal units i and j are designated polygons/locations with correlated attribute values, then $c_{ij}=1$; otherwise, $c_{ij}=0$. Frequently, matrix **C** is converted to its often asymmetric row-standardized counterpart, matrix **W**, for which $w_{ij} = c_{ij}/\sum_{j=1}^{n} c_{ij}$; $\sum_{j=1}^{n} w_{ij} = 1$.

Yet another specification involves inverse distance (i.e., power or negative exponential) between polygon centroids or other points of privilege (e.g., administrative centers, such as capital cities or county seats) within areal unit polygons; these interpoint distances almost always are standardized (i.e., converted to matrix **W**), perhaps with a carefully chosen power or exponent parameter that essentially equates them to their shared common boundary topological structure counterpart. Tiefelsdorf, Griffith, and Boots (1999) discuss other schemes defining a SWM that lies between matrices **C** and **W**.

These nearest neighbor and distance-based specifications allow spatial researchers to posit geographic relationship structures for nonpolygon point observations. By generating Thiessen polygon surface partitionings, these researchers also can posit geographic relationships based upon common boundary rules. Eigenvalues of matrices **C** and **W**, a topic treated in a number of ensuing sections, furnish a quantitative gauge for comparing competing SWMs.

The purpose of a SWM is to define the set of directly correlated observations within a RV, enabling the quantification of SA for a georeferenced attribute. It captures the geometric arrangement of attribute values on a map, often in topological terms.

1.1.3 Different measurements for different data types: Quantifying SA

Similar to correlation coefficients in classical statistics, a SA index may be specific to an attribute's measurement scale (i.e., nominal, ordinal, interval, and ratio). Similar to the Pearson product moment correlation coefficient, r, the Moran coefficient (MC), the most widely employed SA index, can be used with all measurement scales (Griffith, 2010). The MC, originally designed for interval/ratio data, may be defined as follows:

$$\frac{\dfrac{\sum_{i=1}^{n}\sum_{j=1}^{n} c_{ij}(y_i-\bar{y})(y_j-\bar{y})}{\sum_{i=1}^{n}\sum_{j=1}^{n} c_{ij}}}{\sum_{i=1}^{n}(y_i-\bar{y})^2/n} = \frac{n}{\sum_{i=1}^{n}\sum_{j=1}^{n} c_{ij}} \frac{\sum_{i=1}^{n}(y_i-\bar{y})\left[\sum_{j=1}^{n} c_{ij}(y_j-\bar{y})\right]}{(n-1)s^2} = \frac{n}{\sum_{i=1}^{n}\sum_{j=1}^{n} c_{ij}} \frac{\sum_{i=1}^{n} z_i\left(\sum_{j=1}^{n} c_{ij}z_j\right)}{(n-1)}, \tag{1.1}$$

where c_{ij} is an entry in the SWM **C**, $\overline{y}$ is the arithmetic mean and s^2 is the sample variance of response variable Y, and z_i is the z-score for attribute value y_i. The numerator of the MC contains the pairs of values z_i and $\sum_{j=1}^{n} c_{ij}z_j$, whose graphic portrayal is the Moran scatterplot (see Section 1.2.2). Like r, the MC is a covariation-based index. Unlike r—whose extremes are −1 (a perfect indirect relationship) and one (a perfect direct relationship), and for which zero denotes no correlation—the MC's extreme values essentially are a function of the smallest and second largest eigenvalues (a topic treated in ensuing sections) of the employed SWM, and for which $-1/(n-1)$ denotes no SA for a single RV. Frequently, the smallest possible MC is closer to −0.5, whereas the largest possible MC is closer to 1.15. For example, the extreme MCs for the 254-county Texas SWM based upon a covariation perspective are −0.63547 and 1.09798, and based upon a paired comparisons perspective are −0.43024 and 0.89546 (see de Jong, Sprenger, & van Veen, 1984).

The Geary ratio (GR) is a second popular SA index formulated for interval/ratio data. Rather than being based upon cross-products (i.e., covariation), its basis is paired comparisons, or squared differences between those pairs of attribute values whose corresponding row and column entries in the SWM are a positive value. The GR may be defined as follows:

$$\frac{\sum_{i=1}^{n}\sum_{j=1}^{n} c_{ij}\left(y_i - y_j\right)^2 / \sum_{i=1}^{n}\sum_{j=1}^{n} c_{ij}}{2\sum_{i=1}^{n}(y_i-\overline{y})^2/(n-1)} = \frac{n-1}{\sum_{i=1}^{n}\sum_{j=1}^{n} c_{ij}} \frac{\sum_{i=1}^{n}\left(\sum_{j=1}^{n} c_{ij}\right)(y_i-\overline{y})^2}{\sum_{i=1}^{n}(y_i-\overline{y})^2} - \frac{n-1}{n}\text{MC}. \tag{1.2}$$

The right-hand expression emphasizes the negative relationship between the GR and the MC. It also highlights that a GR index value is impacted more than its corresponding MC value by either areal units with relatively large numbers of neighbors (i.e., $\sum_{j=1}^{n} c_{ij}$) or outliers [i.e., $(y_i-\overline{y})^2$]. Here the paired comparison terms, $(y_i - y_j)^2$, indicate that GR index values toward zero represent PSA (i.e., similar values are geographic neighbors); but all of these differences cannot exactly equal zero because then Y would be a constant rather than a RV. A GR index value of one denotes randomness; this expected value is derived by analyzing all possible permutations of a given set of numbers across n locations or by repeatedly drawing random samples from a normal distribution for each of n locations (Cliff & Ord,

1981, p. 21). Larger GR index values represent NSA. Similar to the MC, no definite upper bound exists for the GR. Often its maximum is 2 or more and can be a value in the 100s. Its extreme values also essentially are a function of the second largest and smallest eigenvalues of the employed SWM. Based upon the MC SWM conceptualization, the extreme GR values for the 254-county Texas SWM are 0.08730 and 1.51400, whereas the GR SWM conceptualization yields extreme GR values of 0.00730 and 2.00131.

Replacing the interval/ratio measurement scale with an ordinal measurement scale involving distinct rankings results in

$$MC = \frac{12n}{\sum_{i=1}^{n}\sum_{j=1}^{n} c_{ij}} \frac{\sum_{i=1}^{n}\sum_{j=1}^{n} c_{ij}[y_i - (n+1)/2]\left[y_j - (n+1)/2\right]}{(n^2 - 1)}, \text{and}$$

$$GR = \frac{6n}{\sum_{i=1}^{n}\sum_{j=1}^{n} c_{ij}} \frac{\sum_{i=1}^{n}\sum_{j=1}^{n} c_{ij}\left(y_i - y_j\right)^2}{(n^2 - 1)}.$$

Sen and Sööt (1977) present a study about this type of formulation.

Finally, replacing the interval/ratio measurement scale with a nominal measurement scale, say the commonly used binary zero-one (respectively denoted by Z and O) indicator variable for the presence and absence of a phenomenon, results in

$$MC = \frac{2n}{\sum_{i=1}^{n}\sum_{j=1}^{n} c_{ij}}\left[\frac{ZZ}{n_Z} + \frac{OO}{n - n_Z}\right] - 1, \text{and}$$

$$GR = \frac{n(n-1)}{\sum_{i=1}^{n}\sum_{j=1}^{n} c_{ij}} \frac{ZO}{n_Z(n - n_Z)},$$

where n_z denotes the number of areal units with a measure of zero, and ZZ, OO, and ZO are the standard join count statistics. This formulation underscores differences between the MC and the GR: the MC emphasizes PSA (Z geographically adjacent to Z, and O spatially adjacent to O), whereas the GR emphasizes NSA (Z geographically adjacent to O).

A principal difference between conventional correlation coefficients, such as r, Spearman's rank, and phi, is that the measurement scale of the analyzed variable affects their sampling variance. The SA indices experience this same outcome. The GR is a linear combination of the MC and possibly anomalous data features. The join count statistics relate to the MC and the GR through scaling factors. These various components affect the variance. Consequently, because the MC has the best overall variance properties, it tends to be the preferred SA index by spatial researchers.

1.1.4 The MC: Distributional theory

Distributional theory refers to the sampling distribution of a statistic, which a researcher needs for hypothesis-testing purposes. Here the common null hypothesis is zero SA. Cliff and Ord (1981) present this distributional theory for both randomly sampling an attribute from a normal RV (i.e., the normality assumption) and randomization of a given set of attribute values. They extend this former case to linear regression error terms estimated with ordinary least squares (OLS).

For the case of a single RV, the expected value of the sampling distribution of the MC for either the random sampling or randomization inferential basis is $-1/(n-1)$, signifying zero SA. The asymptotic standard error for this case is $\sqrt{\frac{2}{\sum_{i=1}^{n}\sum_{j=1}^{n} c_{ij}}}$; this asymptotic result is extremely good by $n > 25$ when no covariates are included in an analysis. As n goes to infinity, the sampling distribution of MC converges on a normal distribution. Accordingly, the test statistic is given by

$$z = \sqrt{\sum_{i=1}^{n}\sum_{j=1}^{n} c_{ij}}\,\frac{MC + \frac{1}{n-1}}{\sqrt{2}}. \tag{1.3}$$

Although the most natural alternate hypothesis is that SA is not equal to zero (i.e., a two-tailed test), because almost all geographic phenomena exhibit PSA, spatial researchers almost always could argue for an alternate hypothesis of PSA.

For linear regression residuals, the variance approximation remains useful when the number of covariates is not a function of n and the spatial structure is far from a geographic maximum connectivity case (these two characteristics generally hold in practice; see Appendix 1.A). Similarly, although the expected value is a function of the covariates included in a linear regression equation specification, it tends to converge to zero from a negative value as n increases, given that the number of covariates remains the same.

1.2 Impacts of SA on attribute statistical distributions

The principal impact of SA is on the variance of a RV: PSA inflates variance. Table 1.1 summarizes results from a simulation experiment in which the spatially autocorrelated RVs contain moderate PSA (approximately, MC = 0.7 and GR = 0.3). The four RVs represent the ones most commonly employed in spatial analyses. The mean does not change, whereas the variance substantially increases (e.g., a variance inflation factor of 1.2^2 to 3.0^2) once SA is embedded in a RV. Skewness (i.e., symmetry) tends to be impacted less than variance, with the Poisson RV indicating that existing skewness can be exacerbated by the presence of PSA. Finally, kurtosis (peakedness) tends to be noticeably impacted by the presence of PSA. These summary statistics reveal that the general effect of PSA is to shrink the frequencies of the more central values of a RV and to inflate the frequencies of the values located away from that RV distribution's center (e.g., arithmetic mean): a more platykurtic distribution with fatter tails.

SA is a two-dimensional concept. As such, visualizing it helps to understand it. The relevant systematic organization of georeferenced attribute values is a map pattern. A variety of tools exist that highlight map patterns associated with SA. An obvious one is a map. Fig. 1.1 presents five map patterns depicting different natures and degrees of SA. With regard to the preceding discussion of variance inflation, Fig. 1.1A implies that marked PSA decreases within regions variation as well as increases between regions variation. Fig. 1.1E implies that marked NSA increases within regions variation as well as decreases between regions variation.

Fig. 1.2 presents an example of the variance inflation introduced into a normal RV by PSA. Fig. 1.2A is the histogram for independent and identically distributed (IID) random observations. Its range is roughly −3 to 3. Fig. 1.2B is the histogram for these same data after embedding PSA in them. Its range is roughly −5 to 5. The highest bar in Fig. 1.2A is about 35%, whereas the highest bar in Fig. 1.2B is about 23%. The tails in Fig. 1.2B are much heavier than those in Fig. 1.2A. Nevertheless, both frequency distributions center on zero, and both are reasonably symmetric.

1.2.1 Effects of spatial dependence: Deviating from independent observations

Independent observations is a critical assumption in classical statistics; it indicates that the statistical behavior of one value does not depend upon other values of a RV. It allows the multiplication of probabilities when analyzing

Table 1.1 Averages of 10,000 replications for a simulation based upon a 22-by-23 (i.e., n = 506) regular square tessellation, a rook adjacency definition, and a moderate degree of PSA

Random variable	Mean	Standard deviation		Skewness		Excess kurtosis	
	Random & SA	Random	SA	Random	SA	Random	SA
Normal (0, 1)	0.0003	0.9996	1.7331	−0.0000	−0.0001	0.0001	−0.2582
Bernoulli (p = 0.5)	0.4997	0.5000	1.5013	0.0012	0.0005	−2.0002	−0.4887
Binomial (n = 10, p = 0.5)	4.9971	1.5797	2.1209	−0.0011	−0.0011	−0.1977	−0.1771
Poisson (μ = 5)	5.0002	2.2333	2.6442	0.4426	0.2656	0.1912	0.0504

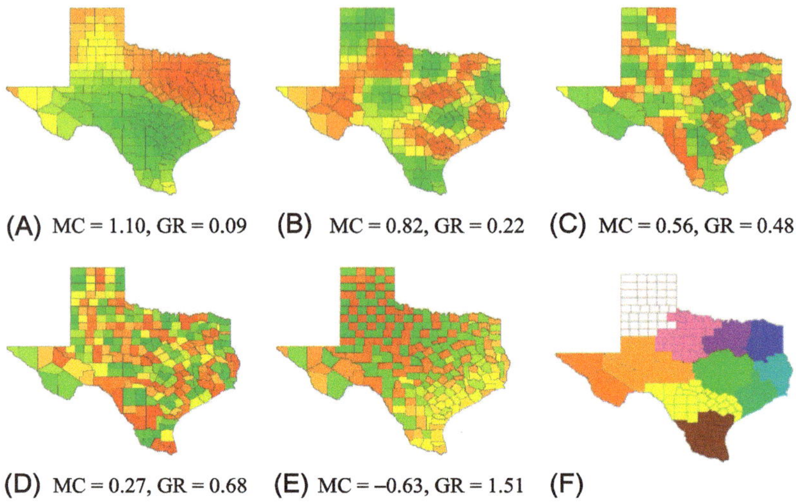

Fig. 1.1 Map patterns portraying selected natures and degrees of SA—the measurement scale goes from red to yellow to green—and a statewide regionalization scheme—the distinct colors denote county membership in a region. (A) Maximum positive. (B) Strong positive. (C) Moderate positive. (D) Weak positive. (E) Maximum negative. (F) Public health regions.

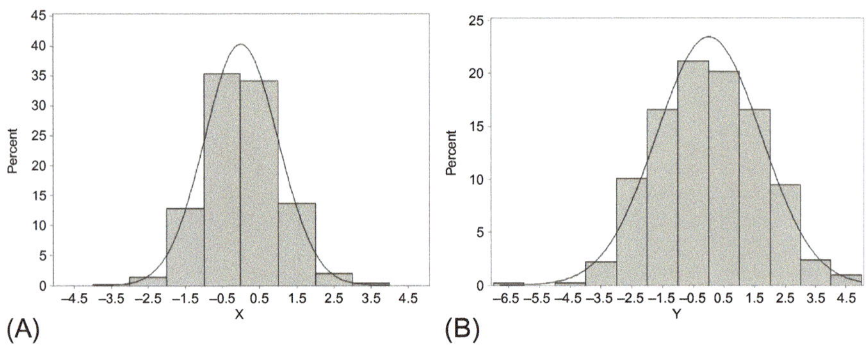

Fig. 1.2 A specimen normal random variable, n = 506. (A) Independent and identically distributed values. (B) Positively spatially autocorrelated values.

joint distributions, reducing many cumbersome cross-product terms to zero, hence dramatically simplifying certain calculations. The presence of nonzero SA violates this assumption. Empirical examples presented in this section demonstrate three different ways PSA tends to impact variance.

Fig. 1.2 illustrates that spatial dependence affects frequency distributions. This topic is treated in more detail in Section 1.2.3. Another impact is upon variance diagnostics. The Poisson example included in Table 1.1 illustrates this notion: theoretically, the mean and the variance are equal, as exemplified by the random data result; PSA increases the variance from 5 to 7, with this overdispersion suggesting that a negative binomial probability model would describe these simulated Poisson data better.

The 11 public health regions of Texas (see http://dshs.texas.gov/chs/info/info_txco.shtm; Fig. 1.1F) furnish a scheme that can illustrate SA impacts on variance across a regionalized geographic landscape. In Table 1.2, the magnitude of the analysis of variance (ANOVA) sums of squares (SS) directly relates to the latent level of SA. The presence of nonzero SA also corrupts the variance across the Texas geographic landscape, resulting in a highly heterogeneous variance (another hallmark of SA), which is indexed here by the Levene (1960) diagnostic statistic. Griffith (1978) discusses issues arising in ANOVA from the presence of nonzero SA. Employing Welch's (1951) ANOVA technique to adjust for nonconstant variance does not necessarily correct this situation. For example, the map data for Fig. 1.1A have their F-ratio decrease to 49.12 (the F-ratios for the other Fig. 1.1 map data, respectively, change to 3.58, 2.16, 0.90, and 0.16). The random map (which is a random permutation of Fig. 1.1A map values) yields the correct type of IID result, with an F-ratio near one.

Calculating the percentage of variance accounted for in a georeferenced response variable by its neighboring values (i.e., redundant locational information) is another way to quantify variance inflation. The geographic distribution of population density (i.e., the ratio of total population to land area) across Texas (Fig. 1.3A) exhibits conspicuous map pattern, with PSA (portrayed in Fig. 1.3B) accounting for roughly 68% of its geographic variation (MC = 0.57 and GR = 0.55, implying moderate PSA).[1] The SA map closely aligns with (i.e., duplicates) the observed population density map. The Austin, Dallas, and Houston metropolitan areas are conspicuous visually in both maps.

The SA map (Fig. 1.3B) portrays the spatially structured information that meaningful covariates may contain. Covariates also can contain spatially unstructured information (e.g., here 32% of the variance in population density across Texas remains unaccounted for). In practice, once a regression equation specification includes meaningful covariates, SA tends to account for an additional 10% of the variance in a georeferenced response variable.

[1] The SA map was constructed with a negative binomial regression whose response variable was total population counts and that included log-area as an offset variable.

Table 1.2 Analysis of variance (ANOVA) results for Fig. 1.1 maps

	ANOVA sums of squares (SS)		Difference of means		Difference of variances	
Map	**Between region**	**Within regions**	**F-ratio**	**Probability**	**Levene**	**Probability**
Fig. 1.1A	0.6996	0.3004	56.59	<0.0001	16.69	<0.0001
Fig. 1.1B	0.0886	0.9114	2.36	0.011	3.63	<0.0001
Fig. 1.1C	0.0267	0.9733	0.67	0.755	3.16	0.001
Fig. 1.1D	0.0216	0.9784	0.54	0.863	2.92	0.002
Random	0.0378	0.9622	0.95	0.484	1.03	0.418
Fig. 1.1E	0.0004	0.9996	0.01	1.000	12.25	<0.0001

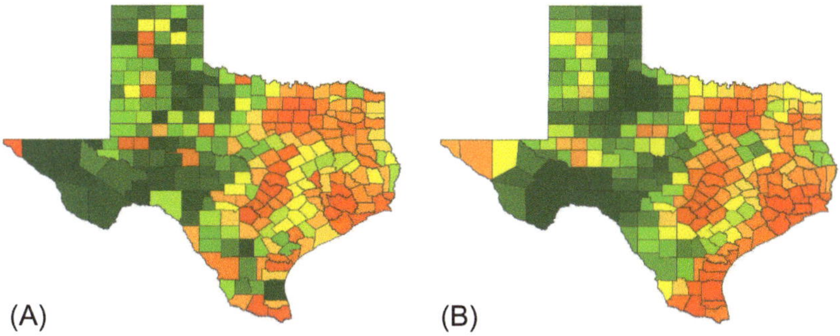

Fig. 1.3 The 2014 population density across Texas by county. (A) Its geographic distribution. (B) Its PSA component.

The sampling distribution of correlation coefficients for georeferenced RVs furnishes a third useful illustration of how PSA affects variance. Consider the following three RVs: log-population density, median household income, and labor force unemployment percentage. Each of these RVs is approximately normally distribution (Fig. 1.4). The raw correlation between income and log-population density is 0.259, whereas it is −0.519 between income and unemployment. In contrast, the raw correlation between the aspatial components (i.e., after these variables are adjusted for the presence of nonzero SA) of income and log-population density is −0.001, whereas the raw correlation between the aspatial components of income and unemployment is −0.334. Moreover, the weak positive relationship between income and log-population density observed across the 254 counties of Texas is solely attributable to SA. The observed negative relationship between income and percentage unemployment is not solely an artifact of latent SA. Instead, SA inflates it, distorting it to be weak rather than moderate in degree.

Fig. 1.5 portrays components (Chapter 3 presents the technique for calculating these quantities) of the geographic distributions of log-population density and median household income. The aspatial components (which, respectively, account for 40% and 56% of the geographic variance of their RVs) appear random:[2] MC = −0.03 and GR = 0.94 for log-population density, and MC = −0.11 and GR = 1.12 for median household income. The common SA components (which, respectively, account for 32% and 26% of the geographic variance of their RVs) display conspicuous map patterns that are quite similar: MC = 0.92 and GR = 0.19 for log-population density, and MC = 0.86 and GR = 0.22 for median household income.

[2] The zero SA value for the MC is negative, asymptotically converging on zero from below.

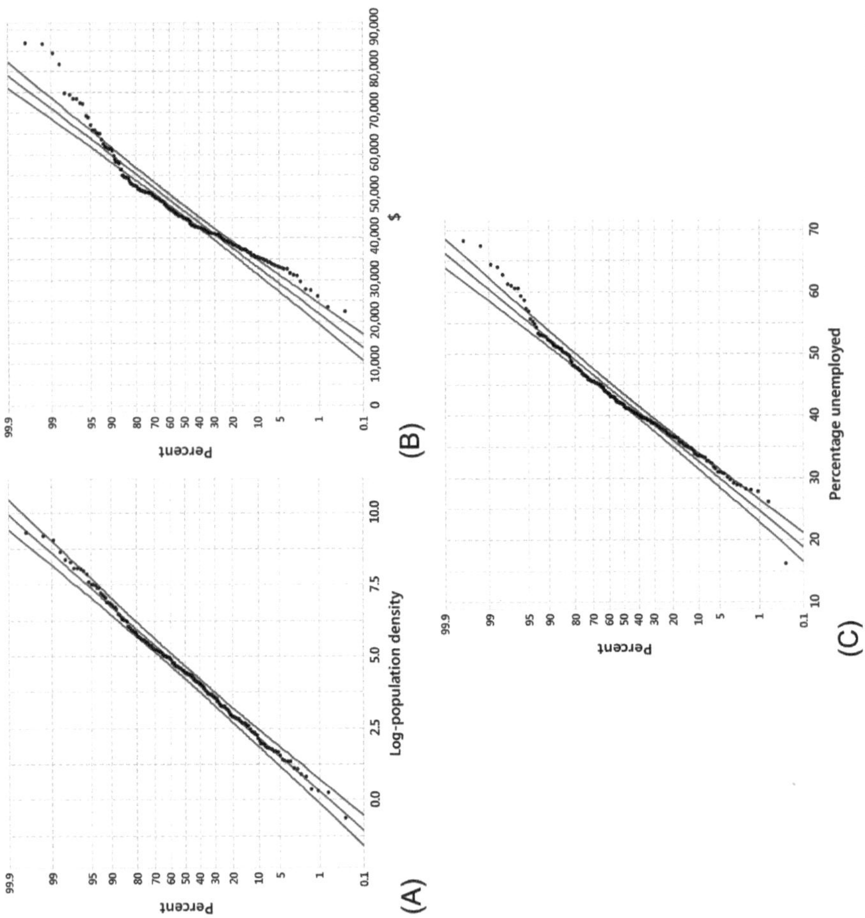

Fig. 1.4 Normal quantile plots for selected random variables. (A) 2014 log-population density. (B) 2014 median household income. (C) 2014 labor force percentage unemployed.

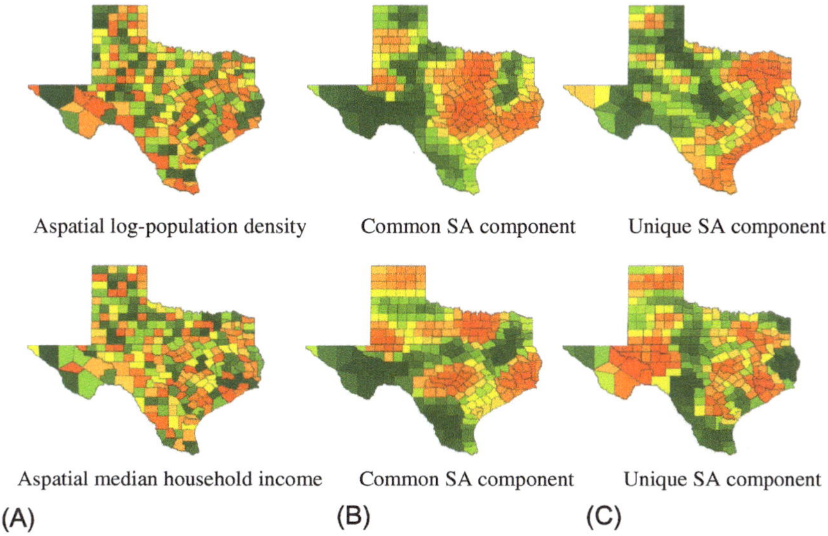

Fig. 1.5 The geographic distribution of 2014 log-population density and median household income. (A) Aspatial components. (B) Common map patterns. (C) Unique map patterns.

These components of SA tend to inflate the observed correlation between the two attribute RVs. The unique SA components (which, respectively, account for 28% and 27% of the geographic variance of their RVs) display conspicuous map patterns that are quite different: MC = 0.86 and GR = 0.20 for log-population density, and MC = 0.77 and GR = 0.29 for median household income. These components of SA tend to deflate the observed correlation between the two attribute RVs. A combination of these inflation and deflation factors alters the sampling distribution variance of a correlation coefficient calculated with two georeferenced RVs.

Table 1.3 summarizes results from a simulation experiment that demonstrates the impact upon the correlation coefficient sampling distribution for the RVs represented by Fig. 1.5. The standard error of the aspatial correlation coefficient is $1/\sqrt{254-1} \approx 0.063$, whereas the conventional standard error calculation for its observed coefficient counterpart is $(1-0.259^2)/\sqrt{254-1} \approx 0.059$. The common SA components alone introduce the most similarity between the two variables, reducing the standard error from 0.63 to 0.59. The unique SA components alone make the two variables more dissimilar, partly compensating for the variance suppression attributable to SA. The total SA (i.e., the combination of common and unique SA components) inflation of the correlation coefficient reduces the sampling variability.

Table 1.3 Bivariate correlation simulation experiment results (10,000 replications)

	SA components			
Statistic	**None**	**Common**	**Unique**	**Common and unique**
Mean	0.000	0.272	–0.000	0.197
Simulated standard error	0.063	0.059	0.057	0.053
IID standard error	0.063	0.061	0.063	0.062
Normal diagnostic probability (K–S)	>0.15	<0.01	0.04	>0.15

Note: K–S denotes the Kolmogorov–Smirnov diagnostic statistic.

A parallel analysis can be conducted with the other pair of variables, namely, median household income and percentage unemployed. Their relationship is addressed briefly here to emphasize that SA is not the only source of association between two variables. SA represents a spatially structured source. Attribute variables also can have a spatially unstructured source. Nevertheless, the important feature is that SA affects the variance of the sampling distribution of statistics.

1.2.2 SA and the Moran scatterplot

The right-hand side of Eq. (1.1) highlights a similarity between the MC and bivariate regression. Moreover, the MC is a regression slope coefficient. If RV Y is converted to its z-score counterpart, yielding an n-by-1 vector of response values **Z**, then the regression coefficient for a no-intercept specification involving regressing vector **CZ** on vector **Z** is given by

$$\left(\mathbf{Z}^{T}\mathbf{Z}\right)^{-1}\mathbf{Z}^{T}\mathbf{C}\mathbf{Z}=\frac{\sum_{i=1}^{n} z_i\left(\sum_{j=1}^{n} c_{ij}z_j\right)}{(n-1)}, \tag{1.4}$$

where superscript T denotes the matrix transpose operation, and $(\mathbf{Z}^{T}\mathbf{Z})^{-1}=(n-1)$. Given the intercept covariate n-by-1 vector of ones, **1**, a second no-intercept regression of vector **C1** on vector **1** yields the following regression slope coefficient:

$$\left(\mathbf{1}^{T}\mathbf{1}\right)^{-1}\mathbf{1}^{T}\mathbf{C}\mathbf{1}=\sum_{i=1}^{n}\sum_{j=1}^{n} c_{ij}/n. \tag{1.5}$$

Eq. (1.1) equals Eq. (1.4) divided by Eq. (1.5).

The bivariate regression associated with Eq. (1.4) also indicates that a scatterplot can be constructed to visualize the SA quantified by a MC. This graphic is a Moran scatterplot. Rather than show two attribute variables, this type of scatterplot portrays the relationship between values on a map and the sum of their neighboring values. Paralleling the interpretation of a conventional scatterplot, if points $\left(z_i, \sum_{j=1}^{n} c_{ij} z_j\right)$ concentrate in the first and third quadrants of the graph, then the nature of the relationship is direct, and the scatter of points depicts PSA: map values tend to be similar to their neighboring map values. As the scatter of points increasingly aligns with the regression coefficient given by expression (1.4), the degree of PSA increases. Fig. 1.1A–D relate to this situation, as do Fig. 1.5B–C. Fig. 1.6A and 1.6B portray Moran scatterplots for, respectively, Fig. 1.1A (which represents maximum PSA) and Fig. 1.3A (i.e., log-population density, which represents moderate PSA).

If points $\left(z_i, \sum_{j=1}^{n} c_{ij} z_j\right)$ concentrate in the second and fourth quadrants of the graph, then the nature of the relationship is indirect, and the scatter of points depicts NSA: map values tend to be dissimilar to their neighboring map values. As the scatter of points increasingly aligns with the regression coefficient given by expression (1.4), the degree of NSA increases. Fig. 1.6C portrays the Moran scatterplot for a modification of Fig. 1.1E (which represents maximum NSA; $MC = -0.50$ and $GR = 1.41$).

Moran scatterplots coupled with maps are effective tools for portraying the nature and degree of SA. As suggested by the preceding regression linkage discussion, they directly relate to spatial autoregression. In contrast, the geostatistical semivariogram plot is a useful tool for portraying SA. It relates to the GR.

1.2.3 SA and histograms

Griffith (2011) outlines impacts of PSA on histograms. Basically, SA acts like a second RV in a parametric mixture model. More specifically, a spatially autocorrelated normal RV resembles a normal-normal mixture model, where the mean is a normal RV. A spatially autocorrelated Poisson RV resembles a Poisson-gamma mixture model, where the mean is a gamma RV. This directly links it to a negative binomial RV. Finally, a spatially autocorrelated binomial RV resembles a beta-binomial mixture model, where the mean is a beta RV. In all three cases, PSA does not affect the mean while it inflates the variance.

Fig. 1.7 portrays impacts on a normal, a Poisson, and a binomial RV for a range of PSA. In all three cases, the principal impact is variance inflation: as

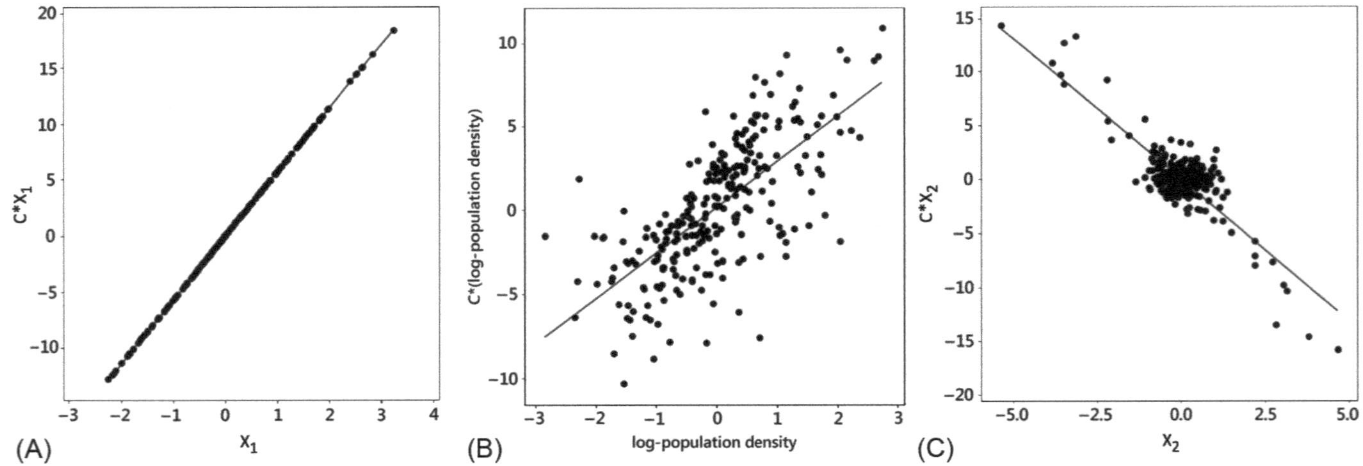

Fig. 1.6 Selected Moran scatterplots constructed with 2014 Texas data. (A) For the Fig. 1.1A map. (B) Log-population density. (C) For a modification of the Fig. 1.1E map.

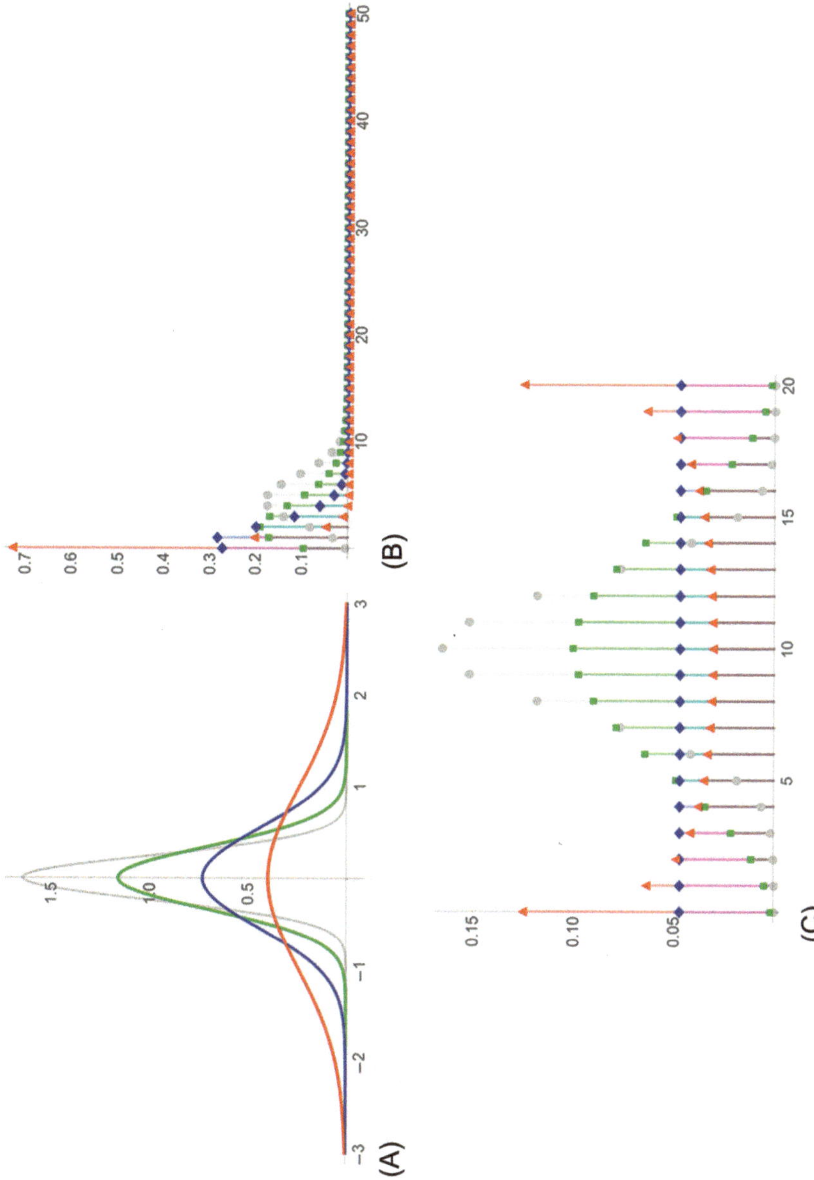

Fig. 1.7 Illustrative impacts of PSA on histograms: gray denotes IID, green denotes weak SA, blue denotes moderate SA, and red denotes strong SA. (A) Normal RV ($\mu=0$, $\sigma=1$). (B) Poisson RV ($\mu=5$). (C) Binomial RV ($p=½$, $n=20$).

the degree of PSA increases, the variance of a RV tends to increase. Accordingly, a normal distribution becomes flatter (this behavior mimics a t-distribution), a Poisson distribution has an increasing number of zeros and outliers (i.e., values farther from the mean), and a binomial distribution moves first toward a discrete uniform distribution form and then to a sinusoidal distribution form (verging on an all-or-nothing situation). In all three cases, the mean is unaffected. Therefore Fig. 1.7 highlights that SA affects not only variance but also kurtosis.

1.3 Summary

This chapter presents findings supporting the motivating contention that recognizing and efficiently and effectively accounting for and handling SA matters in georeferenced data analyses. Its discussion begins with a treatment of various subtly different definitional perspectives about SA, emphasizing this concept in terms of map pattern. A mathematical formularization of Tobler's First Law of Geography follows defining SA. A critical element of quantifying SA is specification of a SWM, whose two most popular forms are denoted by matrix **C**, defined in terms of n binary zero-one indicators based upon the topological structure of a set of areal units, and matrix **W**, the row-standardized version of matrix **C**. A SWM enables the positing of different indices to quantify SA that are specific to different data measurement scale types. Of these indices, the most popular is the MC, which was developed for interval/ratio data measurement scales but has been extended to nominal and ordinal data measurement scales. Accordingly, this chapter presents the basics of the MC sampling distribution theory, which is essential for statistical inference purposes. The next section (Section 1.2) addresses impacts of SA on statistical analyses, stressing various ways it affects variance. Then discussion turns to SA contextualized as a violation of the IID assumption of classical statistics, and how it can be visualized with maps and with a Moran scatterplot, as well as with other graphical tools. Although the SA literature focuses on its impact upon variance, the last substantive/pedagogic background section furnishes illustrations showing how SA more generally can modify the shape of histograms.

Appendix 1.A The mean and variance of the MC for linear regression residuals

Following Cliff and Ord (1973), Upton and Fingleton (1985, p. 338) show that the mean and variance of the MC for linear regression residuals, for which the SWM **C** is symmetric and the number of covariates is p, are as follows, where $\mathbf{M}=\mathbf{I}-\mathbf{X}(\mathbf{X}^{T}\mathbf{X})^{-1}\mathbf{X}^{T}$ is the standard projection matrix from linear regression theory:

$$E(MC)=\frac{n}{\mathbf{1}^{T}\mathbf{C}\mathbf{1}}\frac{TR\left\{\left[\mathbf{I}-\mathbf{X}(\mathbf{X}^{T}\mathbf{X})^{-1}\mathbf{X}^{T}\right]\mathbf{C}\right\}}{n-k}, \text{ and}$$

$$VAR(MC)=2\left(\frac{n}{\mathbf{1}^{T}\mathbf{C}\mathbf{1}}\right)^{2}\left[\frac{TR\left\{\left[\mathbf{I}-\mathbf{X}(\mathbf{X}^{T}\mathbf{X})^{-1}\mathbf{X}^{T}\right]\mathbf{C}\left[\mathbf{I}-\mathbf{X}(\mathbf{X}^{T}\mathbf{X})^{-1}\mathbf{X}^{T}\right]\mathbf{C}\right\}}{(n-k)(n-k+2)} - \frac{\left(TR\left\{\left[\mathbf{I}-\mathbf{X}(\mathbf{X}^{T}\mathbf{X})^{-1}\mathbf{X}^{T}\right]\mathbf{C}\right\}\right)^{2}}{(n-k)^{2}}\right],$$

where $k=p+1$.

For the following cases, suppose $\mathbf{X}_j$ is proportional to the principal eigenvector of matrix $(\mathbf{I}-\mathbf{1}\mathbf{1}^{T}/n)\mathbf{C}(\mathbf{I}-\mathbf{1}\mathbf{1}^{T}/n)$, whose eigenvalue is λ_1, which results in the maximum degree of PSA (a worse-case scenario). The maximum principal eigenvalue for a planar surface partitioning is such that $\lambda_1 \leq a+\sqrt{2n-b}$, for appropriately calculated real numbers a and b (Griffith & Sone, 1995, p. 170). However, most empirical surface partitionings relate more to a mixture of a regular square and a regular hexagonal tessellation. In addition, most empirical surfaces do not have any areal units whose number of neighbors is a function of n; rather, areal units with rook adjacency-defined neighbors of 30 or more are rare. Accordingly, their principal eigenvalues are much closer to 6 and not a function of n.

Case I: Orthogonal covariates in X (which also contains the vector 1 for the intercept term)

Without loss of generality, convert each covariate X_j to its z-score counterpart, z_j. Then the maximum mean and variance values are as follows

Continued

Table 1.A.1 The standard error of the MC calculated with linear regression residuals

Tessellation	Asymptotic	No covariates	two orthogonal covariates	Two perfectly correlated covariates
n = 100				
Linear	0.10050	0.09898	0.09445	0.09733
Square (10-by-10, rook)	0.07454	0.07242	0.06475	0.06953
Hexagonal (10-by-10)	0.06190	0.05925	0.04857	0.05519
Constrained maximum hexagonal (10-by-10)	0.05842	0.05542	0.04423	0.05098
Maximum connectivity	0.05832	0.03243	0.02923	0.03110
n = 400				
Linear	0.05006	0.04987	0.04936	0.04968
Square (20-by-20, rook)	0.03627	0.03601	0.03508	0.03564
Hexagonal (20-by-20)	0.02987	0.02954	0.02830	0.02904
Constrained maximum hexagonal (20-by-20)	0.02916	0.02880	0.02718	0.02812
Maximum connectivity	0.02894	0.01656	0.01621	0.01641
n = 900				
Linear	0.03335	0.03330	0.03315	0.03324
Square (30-by-30, rook)	0.02397	0.02389	0.02362	0.02379
Hexagonal (30-by-30)	0.01968	0.01959	0.01923	0.01944
Constrained maximum hexagonal (30-by-30)	0.01948	0.019038	0.01889	0.01918
Maximum connectivity	0.01927	0.01108	0.01098	0.01104
n = 1,600				
Linear	0.02501	0.02498	0.02492	0.02496
Square (40-by-40, rook)	0.01790	0.01787	0.01776	0.01782
Hexagonal (40-by-40)	0.01468	0.01464	0.01449	0.01458
Constrained maximum hexagonal (40-by-40)	0.01459	0.01455	0.01435	0.01447
Maximum connectivity	0.01444	0.00832	0.00828	0.00830
n = 2,500				
Linear	0.02000	0.01999	0.01996	0.01998
Square (50-by-50, rook)	0.01429	0.01427	0.01421	0.01425
Hexagonal (50-by-50)	0.01170	0.01168	0.01161	0.01165
Constrained maximum hexagonal (50-by-50)	0.01166	0.01164	0.01154	0.01160
Maximum connectivity	0.01155	0.00666	0.00664	0.00665

$$E(MC) = \frac{n}{\mathbf{1}^T\mathbf{C1}} \frac{-\mathbf{1}^T\mathbf{C1}/n - p\lambda_1}{n-k}, \text{ and}$$

$$VAR(MC) \Rightarrow \frac{TR\left\{\left[\mathbf{I} - \mathbf{z}(\mathbf{z}^T\mathbf{z})^{-1}\mathbf{z}^T\right]\mathbf{C}\left[\mathbf{I} - \mathbf{z}(\mathbf{z}^T\mathbf{z})^{-1}\mathbf{z}^T\right]\mathbf{C}\right\}}{(n-k)(n-k+2)}$$

$$= \frac{\mathbf{1}^T\mathbf{C1} - 2\left[\mathbf{1}^T\mathbf{C}^2\mathbf{1}/n + p\lambda_1^2\right] + \left\{\left[\mathbf{1}^T\mathbf{C1}/n\right]^2 + p\lambda_1^2\right\}}{(n-k)(n-k+2)}$$

$$\therefore \lim_{n\to\infty} VAR(MC) = \frac{2}{\mathbf{1}^T\mathbf{C1}}.$$

■

Case II: Perfect multicollinearity among the covariates in X (which also contains the vector 1 for the intercept term)

If the covariates are perfectly correlated, then matrix **X** can be replaced with its single principal component. Accordingly, $k = p + 1 = 2$, and

$$E(MC) = \frac{n}{\mathbf{1}^T\mathbf{C1}} \frac{-\mathbf{1}^T\mathbf{C1}/n - \lambda_1}{n-2}, \text{ and}$$

$$VAR(MC) \Rightarrow \frac{TR\left\{\left[\mathbf{I} - \mathbf{z}(\mathbf{z}^T\mathbf{z})^{-1}\mathbf{z}^T\right]\mathbf{C}\left[\mathbf{I} - \mathbf{z}(\mathbf{z}^T\mathbf{z})^{-1}\mathbf{z}^T\right]\mathbf{C}\right\}}{n(n-2)}$$

$$= \frac{\mathbf{1}^T\mathbf{C1} - 2\left[\mathbf{1}^T\mathbf{C}^2\mathbf{1}/n + \lambda_1^2\right] + \left\{\left[\mathbf{1}^T\mathbf{C1}/n\right]^2 + \lambda_1^2\right\}}{n(n-2)}$$

$$\therefore \lim_{n\to\infty} VAR(MC) = \frac{2}{\mathbf{1}^T\mathbf{C1}}.$$

■

Case III: Some multicollinearity among the covariates in X (which also contains the vector 1 for the intercept term)

This intermediate case has the same type of results, with p being replaced with some value between 1 and p. ■

Convergence on the asymptotic variance presented here tends to be slower as the number of covariates increases. Nevertheless, it is a good approximation for cases where $n-k$ is of a practical magnitude (e.g., at least 100).

Table 1.A.1 summarizes results for a numerical example in which the number of areal units and the geographic configuration vary. One modification indicated by Case II is that multiple covariates that are correlated are replaced by their principal components, in a dimension-reducing way, which decreases the value of k here in the denominator (an argument could be made not to decrease k). Furthermore, the constrained maximum hexagonal tessellation has areal units labeled 1 and n that wrap around the outside of a complete rectangular region, but constrained so that their maximum numbers of neighbors are 30. Results in this table corroborate the contention that the asymptotic standard error of the MC for regression residuals is a very good approximation as long as the principal eigenvalue of matrix **C** is not a function of n, no areal units have numbers of neighbors that are a function of n, and the number of covariates is not a function of n. Griffith (2010) demonstrates this same result for the case of no covariates. One important index here is $n-k$, which needs to be relatively large for the asymptotic standard error to be a good approximation. The smallest value in Table 1.A.1 for this expression is 97, for which the approximation is quite good. Results for $n<50$ tend to be poor in the linear regression context, whereas results for $n<25$ tend to be poor in the univariate context (i.e., no covariates).

The expected value of the MC is easier to calculate than its standard error; this quantity goes to zero as n goes to infinity.

References

Besag, J. (1974). Spatial interaction and the statistical analysis of lattice systems. *Journal of the Royal Statistical Society: Series B: Methodological, 36*, 192–225.

Cliff, A., & Ord, J. (1973). *Spatial autocorrelation.* London: Pion.

Cliff, A., & Ord, J. (1981). *Spatial processes.* London: Pion.

de Jong, P., Sprenger, C., & van Veen, F. (1984). On extreme values of Moran's I and Geary's c. *Geographical Analysis, 16*, 17–24.

Fingleton, B. (1986). Analyzing cross-classified data with inherent spatial dependence. *Geographical Analysis, 18*, 48–61.

Getis, A. (2008). A history of the concept of spatial autocorrelation: a geographer's perspective. *Geographical Analysis, 40*, 297–309.

Goodchild, M. (1986). Spatial autocorrelation. In *Concepts and techniques in modern geography 47*. Norwich, UK: Geo Books.

Griffith, D. (1978). A spatially adjusted ANOVA model. *Geographical Analysis, 10*, 296–301.

Griffith, D. (1987). *Spatial autocorrelation.* Washington, DC: Association of American Geographers.

Griffith, D. (1992). What is spatial autocorrelation? Reflections on the past 25 years of spatial statistics. *L'Espace Géographique, 21*, 265–280.

Griffith, D. (2005). Spatial autocorrelation. K. Kempf-Leonard (Ed.), 3. *Encyclopedia of social measurement* (pp. 581–590). Amsterdam: Elsevier.

Griffith, D. (2009). Methods: spatial autocorrelation. In R. Kitchin & N. Thrift (Eds.), *International encyclopedia of human geography* (pp. 396–402). New York: Elsevier.

Griffith, D. (2010). The Moran coefficient for non-normal data. *Journal of Statistical Planning and Inference, 140*, 2980–2990.

Griffith, D. (2011). Positive spatial autocorrelation impacts on attribute variable frequency distributions. *Chilean Journal of Statistics, 2*(2), 3–28.

Griffith, D. (2016). Spatial autocorrelation. *Oxford Bibliographies.* https://doi.org/10.1093/OBO/9780199874002-0128.

Griffith, D., & Sone, A. (1995). Trade-offs associated with normalizing constant computational simplifications for estimating spatial statistical models. *Journal of Statistical Computation and Simulation, 51*, 165–183.

Legendre, P. (1993). Spatial autocorrelation: trouble or new paradigm? *Ecology, 74*, 1659–1673.

Levene, H. (1960). Robust tests for equality of variances. In I. Olkin, S. Ghurye, W. Hoeffding, W. Madow, & H. Mann (Eds.), *Contributions to probability and statistics: Essays in honor of Harold Hotelling* (pp. 278–292). Palo Alto, CA: Stanford University Press.

Odland, J. (1988). *Spatial autocorrelation.* Thousand Oaks, CA: SAGE.

Rogerson, P. (1998). A spatial version of the chi-square goodness-of-fit test and its application to tests for spatial clustering. In C. Amrhein, D. Griffith, & J. Huriot (Eds.), *Econometric advances in spatial modelling and methodology, essays in honor of lean Paelinck* (pp. 71–84). Dordrecht: Kluwer.

Sen, A., & Sööt, S. (1977). Rank tests for spatial correlation. *Environment and Planning A, 9*, 897–905.

Teugels, J. (1990). Some representations of the multivariate Bernoulli and binomial distributions. *Journal of Multivariate Analysis, 32*, 256–268.

Tiefelsdorf, M., Griffith, D., & Boots, B. (1999). A variance stabilizing coding scheme for spatial link matrices. *Environment and Planning A, 31*, 165–180.

Tobler. W. (1969). A computer movie simulating urban growth in the Detroit region. Paper prepared for meeting of International Geographical Union, Commission on Quantitative Methods, Ann Arbor, Michigan, August, Published in 1970. *Economic Geography, 46*(2), 234–240.

Upton, G., & Fingleton, B. (1985). *Spatial data analysis by example. 1.* New York: Wiley.

Welch, B. (1951). On the comparison of several mean values: an alternative approach. *Biometrika, 38*, 330–336.

CHAPTER 2

An introduction to spectral analysis

Chapter outline

Spectral analysis appears in a number of different geospatial science contexts. In one, it furnishes a description of a given map pattern, using a combination of sine (SIN) and/or cosine (COS) functions of various amplitude, frequency, phase, period, and power to describe a geographic distribution's finite set of data (see Rayner, 1971). Moreover, cyclical trends furnish a useful summary of map pattern, with SIN and COS functions capturing periodic tendencies in georeferenced data (see Griffith, 1988, Chapter 5). Multispectral remote sensing provides yet another geospatial context, where map pattern in an image is the result of differential light reflectance from a surface.[1] This situation mirrors the literal meaning of the term spectral: an instrument (e.g., a satellite sensor) acquires visible, near-infrared, and shortwave infrared light reflectance signatures in, say, eleven broad wavelength

[1] Measurements are taken in, say, seven spectral bands that quantify the (electromagnetic) spectral reflectance (in micrometers), the fraction of reflected solar energy. Each resulting value is a digital number in the range [0, 255]. For Landsat TM + data, the ground cell geographic resolution is 30 m for Bands 1–6 and 120 m for Band 7.

https://doi.org/10.1016/B978-0-12-815043-6.00002-1

bands for Landsat 8. Table 2.1 summarizes features of these measurements that relate them to their common interpretations.

Fig. 2.1 presents a 4,665,847- and a 4,666,735-pixel (approximately 2209-by-2164) remotely sensed satellite image of Houston, Texas—for a pre- and a post-Hurricane Harvey (August 25, 2017) date—portraying the normalized difference vegetation index (NDVI)[2] and illustrating output

Table 2.1 Ready reference for the meaning of the 30-m resolution spectral signatures

Band	Wavelength	Spectral region	Principal applications	Target reflectance
1	0.435–0.451	Coastal/aerosol	Shallow water; fine particles (e.g., dust, smoke)	Ocean color
2	0.452–0.512	Blue	Water body penetration; forest-type	Soil–vegetation separation
3	0.533–0.590	Green	Vegetation discrimination and vigor assessment	Vegetation
4	0.636–0.673	Red	Plant species differentiation	Chlorophyll absorption
5	0.851–0.879	Near-infrared	Vegetation types, vigor and biomass; soil moisture	Water bodies
6	1.566–1.651	Panchromatic and shortwave infrared 1	Vegetation and soil moisture	Vegetation moisture
7	2.107–2.294	Panchromatic and shortwave infrared 2	Mineral/rock type discrimination	Hydrothermal
9	1.363–1.384	Cirrus	Thin clouds composed of ice crystals	High-altitude cloud contamination

Note: Bands 8, 10, and 11 are not included here because of their different spatial resolutions, and because Bands 10 and 11 combined are equivalent to Landsat 7 Band 6; spectral indices employed here (Table 2.2) do not require any of these three particular bands.

[2] These images are from Landsat 8 OLI/TIRS C1 Level-1 (30 m resolution; available at https://glovis.usgs.gov/app) and are based upon the NDVI calculated with spectral bands 4 and 5: April 6, 2017 and, January 3, 2018. Selection of these two dates was based upon cloud-free images preceding and following Hurricane Harvey's landfall. The Houston city limits were obtained from https://cohgis-mycity.opendata.arcgis.com/datasets/houston-city-limit.

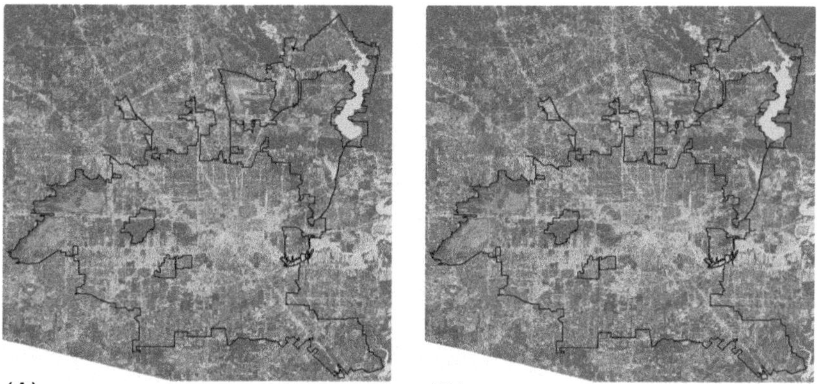

Fig. 2.1 The geographic distribution of the NDVI across the Houston, Texas, urban region (*dark black lines* denote city limits); the NDVI range of −0.28 to 0.64 is directly proportional to the *darkness of the green*. (A) April 6, 2017: pre-Hurricane Harvey. (B) January 3, 2018: post-Hurricane Harvey.

Table 2.2 Selected common measures calculated with Landsat 8 spectral band data

Acronym	Name	Spectral band calculation
ARVI	Atmospherically resistant vegetation index	(Band 5 − 2Band 4 + Band 2)/(Band 5 + 2Band 4 − Band 2)
NDVI	Normalized difference vegetation index	(Band 5 − Band 4)/(Band 5 + Band 4)
OSAVI	Optimized soil adjusted vegetation index	(1.16) (Band 5 − Band 4)/(Band 5 + Band 4 + 0.16)
RVI	Ratio vegetation index	(Band 5)/(Band 4)
TSAVI	Transformed soil adjusted vegetation index	(a×Band 5 − a×Band 4 − b)/(a×Band 5 + Band 4 − a×b)

obtainable from collected multispectral remotely sensed data. Chapter 4 presents a big data example based upon a subregion of this dataset. Table 2.2 summarizes some of the commonly calculated indices that employ remotely sensed spectral band values, of which NDVI is one. Some of the illustrative analyses appearing in Chapter 4 use these data.

In general, spectral analysis involves a spectrum of frequencies, such as the eigenvalues of a spatial weights matrix's (SWM's) eigenfunctions. This chapter treats selected spectral analysis features that are relevant to spatial autocorrelation (SA) and hence Moran eigenvector spatial filtering (MESF).

2.1 Representing SA in the spectral domain

A close relationship exists between SA and the spectral domain (see Bartlett, 1975). A spatial domain refers to how a map pattern changes over space, with this geographic variation relating to SA, whereas a spectral domain refers to how much of the map pattern lies within each given frequency band over a range of frequencies, which constitutes a spectrum. A mathematical function relates these two domains.

Tiefelsdorf and Boots (1995) establish an important link between a n-by-n SWM **C** and SA: the eigenvalues of a modified SWM, namely, the Moran coefficient (MC) numerator matrix expression $(\mathbf{I}-\mathbf{11}^{\mathrm{T}}/\mathrm{n})\times\mathbf{C}(\mathbf{I}-\mathbf{11}^{\mathrm{T}}/\mathrm{n})$ involving pre- and postmultiplication of a SWM by a standard projection matrix,[3] index all possible distinct natures and degrees of SA associated with a given SWM. This set of eigenvalues constitutes a spectrum. Griffith (1996) establishes the accompanying link between SA and map pattern: a linear combination of the eigenvectors of the aforementioned modified SWM describes the map pattern exhibited by a geographic distribution.

2.1.1 SA: From a spatial frequency to a spatial spectral domain

Table 2.1 and Fig. 2.1 present a common interface between spatial analysts and spectral concepts, namely, a remotely sensed satellite image. Here a multisensor instrument records wavelength measurements of electromagnetic radiation, with each discrete, distinctly recorded wavelength interval measured by a sensor constituting a band. In this context, the term spectrum refers to classifying a measurement according to its position on a scale between two extreme or opposite points. This general definition acknowledges that many types of measurements can be labeled spectra. Those of interest in this chapter include SWM eigenvalues and spatial correlations. A glass prism[4] exemplifies the notion of a spectrum specifically with regard to remotely sensed images (Fig. 2.2A).

[3] **I** denotes the n-by-n identity matrix, **1** denotes the n-by-1 vector of ones, and superscript T denotes the matrix transpose operation.

[4] A prism may be defined as a polished, transparent, solid triangle whose two end faces are similar, equal, and parallel triangles, whose three sides are parallelograms, and whose refracting surfaces separate white light—which comprises different colors constituting the set of wavelengths visible to the human eye—into a rainbow of colors.

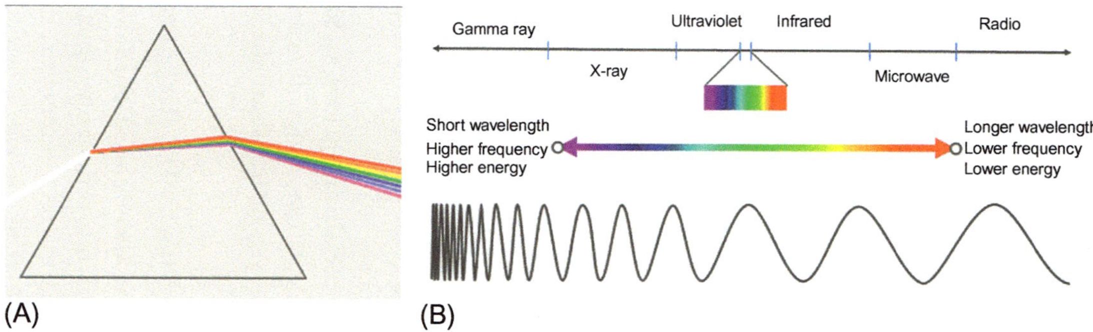

Fig. 2.2 (A) A prism (from Mathematica 11.1.1.0; after Peria, 2007). (B) The electromagnetic radiation spectrum (after imagine.gsfc.nasa.gov/science/toolbox/emspectrum1.html).

The notion of a spectrum implies two distinct reference points for spatial statistical analysis. One is known as the frequency domain, and the other is known as the spectral domain. A spatial frequency domain refers to the frequency with which map attribute values change over a geographic landscape. This is the domain in which MESF operates. It relates to spatial autoregression and hence to inverse spatial covariance matrices constructed with SWMs. The other—the spectral domain—deals with spatial correlation and furnishes selected geographic landscape equations (i.e., spectral density functions) for calculating these correlations. This is the domain in which geostatistics operates. It relates to semivariogram models and hence to spatial covariance matrices (rather than their inverses). SWMs, among other mathematical tools, bridge these two domains (with the matrix inversion operation difference between them magnifying geographic edge effects between these two forms of spatial statistical analysis). This chapter highlights MESF eigenvalues that appear in spectral density functions, emphasizing this linkage.

2.1.2 Eigenvalues and eigenvectors

Multivariate statistical analysis is perhaps the most common subject area in which researchers and students alike first encounter eigenfunctions. In this setting, calculations of these mathematical quantities are for correlation and covariance matrices, among others, and they relate to patterns of multicollinearity. Consequently, that eigenfunctions for SWMs are useful and illuminating should not be a surprise.

Regardless of whether a correlation matrix, a SWM, or some other square matrix is of interest, calculation of eigenfunctions remains the same. Eigenvalues for a given real n-by-n matrix $\mathbf{C}$ are the n roots of the polynomial defined by.

$$\mathrm{Det}(\mathbf{C} - \lambda \mathbf{I}) = 0, \tag{2.1}$$

where Det denotes the matrix determinant operation, and λ denotes an eigenvalue. The solution to this nth-order polynomial is the set of n eigenvalues λ_j, $j = 1, 2, \ldots, n$. If a real matrix is symmetric, then its eigenfunctions are guaranteed to be real numbers, and its eigenvectors are mutually orthogonal. Pairing with each eigenvalue is an n-by-1 eigenvector satisfying the following equation:

$$(\mathbf{C} - \lambda_j \mathbf{I})\mathbf{E}_j = \mathbf{0},\ j = 1, 2, \ldots, n, \tag{2.2}$$

where $\mathbf{E}_j$ denotes the jth eigenvector, and $\mathbf{E}_j \neq \mathbf{0}$ (the trivial solution), with Eq. (2.2) being constrained such that

$$\mathbf{E}_j{}^T\mathbf{E}_j = 1 \text{ (normalization), and } \mathbf{E}_j{}^T\mathbf{E}_k = 0 \text{ (orthogonality)}, j \neq k.$$

Because normalization is a sum-of-squares restriction, it reveals that the eigenvectors are unique except for a multiplicative factor of −1. Although these eigenvectors are orthogonal, they are not uncorrelated. The Perron–Frobenius theorem ensures that the principal eigenvector of SWM **C** has all nonnegative elements. Notably, the matrix $(\mathbf{I}-\mathbf{11}^T/n)$ in the modified matrix expression $(\mathbf{I}-\mathbf{11}^T/n)\mathbf{C}(\mathbf{I}-\mathbf{11}^T/n)$ replaces the principal eigenvector of matrix **C** with one proportional to vector **1** and centers (i.e., forces a mean of zero for) the remaining $n-1$ eigenvectors, resulting in them being mutually uncorrelated as well as mutually orthogonal.

The outcome of Eqs. (2.1) and (2.2) is the following set of n MC values that index the nature and degree of SA in their corresponding eigenvectors:

$$\begin{aligned} MC_j &= \frac{n}{\mathbf{1}^T\mathbf{C}\mathbf{1}} \frac{\mathbf{E}_j^T(\mathbf{I}-\mathbf{11}^T/n)\mathbf{C}(\mathbf{I}-\mathbf{11}^T/n)\mathbf{E}_j}{\mathbf{E}_j^T(\mathbf{I}-\mathbf{11}^T/n)\mathbf{E}_j} = \frac{n}{\mathbf{1}^T\mathbf{C}\mathbf{1}} = \frac{\mathbf{E}_j^T\mathbf{C}\mathbf{E}_j}{\mathbf{E}_j^T\mathbf{E}_j} = \frac{n}{\mathbf{1}^T\mathbf{C}\mathbf{1}}\frac{\lambda_j}{1} \\ &= \lambda_j \frac{n}{\mathbf{1}^T\mathbf{C}\mathbf{1}}; \end{aligned} \tag{2.3}$$

and a set of n eigenvectors that portrays the full range of SA, from maximum positive to maximum negative, for a posited SWM characterizing a given geographic landscape surface partitioning.

2.1.3 Principal components analysis: A reconnaissance

Eigendecomposition (i.e., rewriting a matrix in terms of its eigenvalues and eigenvectors) of a correlation matrix produces a set of eigenvalues that can be arranged in descending order, along with their corresponding eigenvectors. These eigenvectors are referred to as principals components (PCs) and have the intriguing property that the amount of variance they capture follows the same order as their magnitudes. In other words, the first eigenvector summarizes the most variance, the second eigenvector summarizes the second largest amount of variance, and so forth. Because the first few eigenvectors often are sufficient for representing most of the information contained in their correlation matrix, principal components analysis (PCA) commonly is used for variable selection. However, here we use PCA to illustrate linkages between eigenvectors and SA.

Griffith (1984) began to explore the application of PCA to the binary SWM **C**, initially making little headway because the analysis was of n indicator variable correlation (i.e., phi) coefficients. Later, a simple modification of matrix **C** illuminated a breakthrough: adding an identity matrix to **C** [i.e., $(\mathbf{C}+\mathbf{I})$] renders a SWM mimicking a correlation matrix, allowing the application of conventional PCA to it in a way that yields interpretable results (see Appendix 2.A). Furthermore, based on the mathematical properties of eigenvalues and eigenvectors, matrices $(\mathbf{C}+\mathbf{I})$ and **C** have the same eigenvectors and the same eigenvalues if one is added to the eigenvalues of matrix **C** (i.e., λ_j+1). The resulting PCA gives an important insight into the asymptotic relationship between the SWM **C** eigenvectors and SA: the second PC corresponds to the maximum positive SA (PSA), and the last PC to the maximum negative SA (NSA). More specifically, PCA made possible the discovery of, and now makes obvious, the fundamental theorem of MESF, which pertains to matrix $(\mathbf{I}-\mathbf{1}\mathbf{1}^{\mathrm{T}}/\mathrm{n})\mathbf{C}(\mathbf{I}-\mathbf{1}\mathbf{1}^{\mathrm{T}}/\mathrm{n})$:

> *The first eigenvector, say* $\mathbf{E}_1$*, is the set of real numbers that has the largest MC achievable by any such set for the geographic arrangement defined by the spatial connectivity matrix* **C***; the second eigenvector,* $\mathbf{E}_2$*, is the set of real numbers that has the largest achievable MC by any such set that is orthogonal and uncorrelated with* $\mathbf{E}_1$*; the third eigenvector is the third such set of real numbers; and so on through* $\mathbf{E}_n$*, the set of real numbers that has the largest negative MC achievable by any real number set that is orthogonal and uncorrelated with the preceding (n − 1) eigenvectors.*

These findings, along with proof of uncorrelatedness furnished a few years later by Griffith (2000), laid an important conceptual foundation for eigenvector spatial filtering (Tiefelsdorf & Boots, 1995; Griffith, 1996).

2.1.4 The spectral decomposition of a modified SWM

The MC can be written in matrix form as

$$\frac{\mathrm{n}}{\mathbf{1}^{\mathrm{T}}\mathbf{C}\mathbf{1}}\frac{\mathbf{Y}^{\mathrm{T}}(\mathbf{I}-\mathbf{1}\mathbf{1}^{\mathrm{T}}/\mathrm{n})\mathbf{C}(\mathbf{I}-\mathbf{1}\mathbf{1}^{\mathrm{T}}/\mathrm{n})\mathbf{Y}}{\mathbf{Y}^{\mathrm{T}}(\mathbf{I}-\mathbf{1}\mathbf{1}^{\mathrm{T}}/\mathrm{n})\mathbf{Y}}. \tag{2.4}$$

The modified matrix $(\mathbf{I}-\mathbf{1}\mathbf{1}^{\mathrm{T}}/\mathrm{n})\mathbf{C}(\mathbf{I}-\mathbf{1}\mathbf{1}^{\mathrm{T}}/\mathrm{n})$ in its numerator is the matrix that relates to SA. Accordingly, the pair of eigenfunction problems becomes

$$\mathrm{Det}\left[\left(\mathbf{I}-\mathbf{1}\mathbf{1}^{\mathrm{T}}/\mathrm{n}\right)\mathbf{C}\left(\mathbf{I}-\mathbf{1}\mathbf{1}^{\mathrm{T}}/\mathrm{n}\right)-\lambda\mathbf{I}\right]=0,$$

and

$$\left[\left(\mathbf{I}-\mathbf{1}\mathbf{1}^{\mathrm{T}}/\mathrm{n}\right)\mathbf{C}\left(\mathbf{I}-\mathbf{1}\mathbf{1}^{\mathrm{T}}/\mathrm{n}\right)-\lambda_j\mathbf{I}\right]\mathbf{E}_j=\mathbf{0},\mathrm{j}=1,2,\ldots,\mathrm{n},$$

subject to $\mathbf{E}_j^T\mathbf{E}_j=1$ and $\mathbf{E}_j^T\mathbf{E}_k=0$, $j\neq k$. Because matrix $(\mathbf{I}-\mathbf{11}^T/n)\times \mathbf{C}(\mathbf{I}-\mathbf{11}^T/n)$ is square and symmetric, it can be decomposed into a series of rank 1 matrices consisting of the products of the eigenvalues and eigenvectors; that is,

$$\left(\mathbf{I}-\frac{\mathbf{11}^T}{n}\right)\mathbf{C}\left(\mathbf{I}-\frac{\mathbf{11}^T}{n}\right)=\sum_{j=1}^{n}\lambda_j\mathbf{E}_j\mathbf{E}_j^T.$$

This right-hand summation expression is a matrix expansion of the standard eigendecomposition $(\mathbf{I}-\mathbf{11}^T/n)\mathbf{C}(\mathbf{I}-\mathbf{11}^T/n)=\mathbf{E\Lambda E}^T$, where $\mathbf{\Lambda}$ is the diagonal matrix containing the n eigenvalues λ_j.

2.1.5 Representing the MC with eigenfunctions

Let vector $\mathbf{Y}=\mathbf{E}_j$. Next, substitute $\mathbf{E}_j$ into Eq. (2.4):

$$\left[n/\left(\mathbf{1}^T\mathbf{C1}\right)\right]\mathbf{E}_j{}^T\left(\mathbf{I}-\mathbf{11}^T/n\right)\mathbf{C}\left(\mathbf{I}-\mathbf{11}^T/n\right)\mathbf{E}_j/$$
$$\left[\mathbf{E}_j{}^T\left(\mathbf{I}-\mathbf{11}^T/n\right)\mathbf{E}_j\right]=\left[n/\left(\mathbf{1}^T\mathbf{C1}\right)\right]\lambda_j/1,$$

which is the same as Eq. (2.3). This outcome highlights why the extreme eigenvalues of matrix $(\mathbf{I}-\mathbf{11}^T/n)\mathbf{C}(\mathbf{I}-\mathbf{11}^T/n)$, λ_1 and λ_n, respectively, define the maximum and minimum values of a MC—they optimize its Rayleigh quotient. This feature of the MC distinguishes it from a Pearson product moment correlation coefficient: it does not have extreme values of ±1 but rather has an upper bound that often exceeds 1 by as much as 0.15, or more, and a lower bound that often is closer to -0.5 than to -1.

Fig. 2.3A portrays the 2010 census tract surface partitioning ($n=1052$) for the Dallas–Fort Worth (DFW) metroplex, and Fig. 2.3B is ArcMap SA output. The principal eigenvalue for this surface partitioning is 6.2675604, based on the rook's contiguity, with $\mathbf{1}^T\mathbf{C1}=5626$. Therefore for PSA, MC $=(1052/5626)$ $(6.2675604)=1.171965$. This maximum PSA value exceeds 1 by >0.17. Meanwhile, for NSA, MC $=(1052/5626)\times(-3.949034)=-0.738426$. This maximum NSA value is closer to -0.5 than to -1.

2.1.6 Visualizing map patterns with eigenvectors

By convention, a SWM's rows and columns are ordered with the same sequence of areal units. The eigenvectors of a SWM are n-by-1 vectors, with each of their rows linking to the corresponding areal unit identifier (ID) for the same row in their parent SWM. This one-to-one correspondence

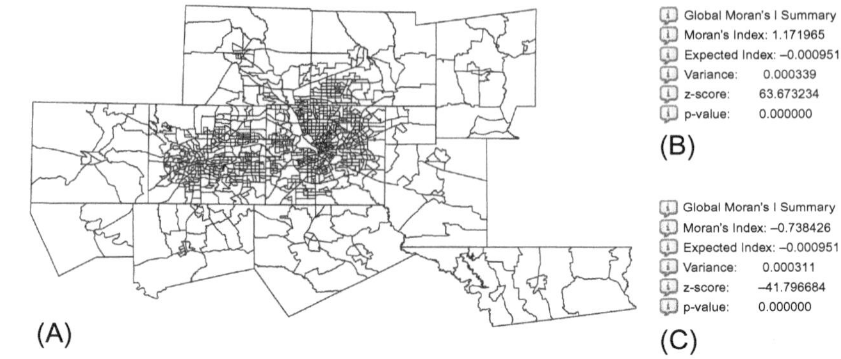

Fig. 2.3 The DFW metropolitan area. (A) The 2010 census tracts surface partitioning. (B) A screenshot of the ESRI ArcMap PSA MC calculation results. (C) A screenshot of the ESRI ArcMap NSA MC calculation results.

enables mapping the eigenvectors of a SWM. The individual eigenvector elements simply have to be joined to their corresponding polygons in a shapefile file for visualization by ArcMap, for example.

Fig. 2.4 portrays the extreme eigenvectors for matrix $(\mathbf{I}-\mathbf{11}^{\mathrm{T}}/n)\times \mathbf{C}(\mathbf{I}-\mathbf{11}^{\mathrm{T}}/n)$ representing the 2010 DFW metroplex census tracts. Fig. 2.4A portrays the principal eigenvector map pattern, the maximum PSA case, which depicts a hill/valley (this would change from a hill to a valley, or vice versa, by multiplying the eigenvector by -1) roughly in the center of the region, surrounded by a trough/hill, with intermediate plateaus in the western and southeastern parts of the region. This hill map pattern is one of the common global geographic trends portrayed by eigenvectors. Another typical map pattern is an east–west gradient trend, adapting to the somewhat rectangular shape of the DFW metroplex. Fig. 2.3A reveals that the hill/valley subregion in Fig. 2.4A coincides with a concentration of numerous smaller census tracts located mostly in Dallas and Tarrant Counties. In contrast, Fig. 2.4B portrays a map pattern with considerable neighboring contrasts (e.g., *dark red* in juxtaposition with *dark green*), an alternating trend that exemplifies NSA. This map pattern has its intermediate values in its periphery, encircling the stark contrasts created by the alternating trend. A comparison of Fig. 2.4A and B reveals that PSA exhibits more smoothness in its map pattern, and NSA exhibits more fragmentation in its map pattern. These tendencies are less conspicuous in visualizations of the less extreme MC values.

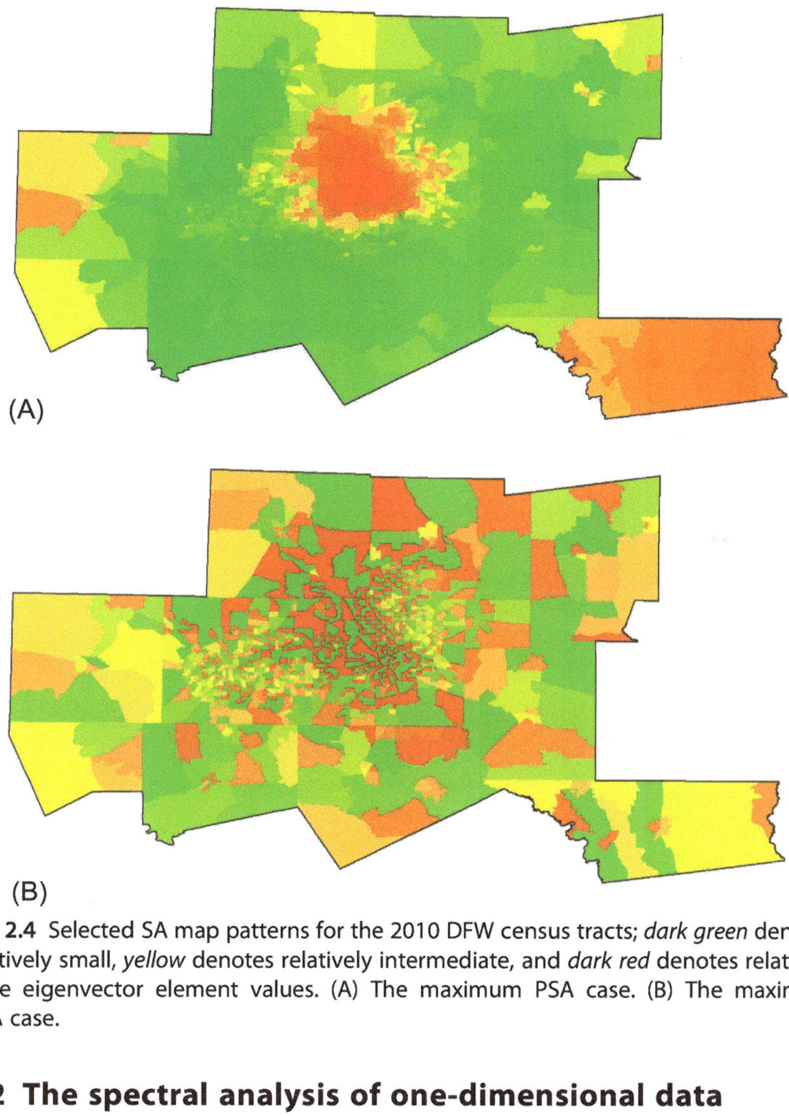

Fig. 2.4 Selected SA map patterns for the 2010 DFW census tracts; *dark green* denotes relatively small, *yellow* denotes relatively intermediate, and *dark red* denotes relatively large eigenvector element values. (A) The maximum PSA case. (B) The maximum NSA case.

2.2 The spectral analysis of one-dimensional data

A simple one-dimensional geographic landscape (e.g., Fig. 2.5A) illustrates the connection between the spatial frequency and spatial spectral domains, furnishing the basis of the spectral density function equations presented in this section. SA indicates how rapidly a map pattern changes across a geographic landscape, using randomness as its yardstick. SA contains information about the expected frequency content mentioned earlier in this chapter. Meanwhile, the spectral density function furnishes one

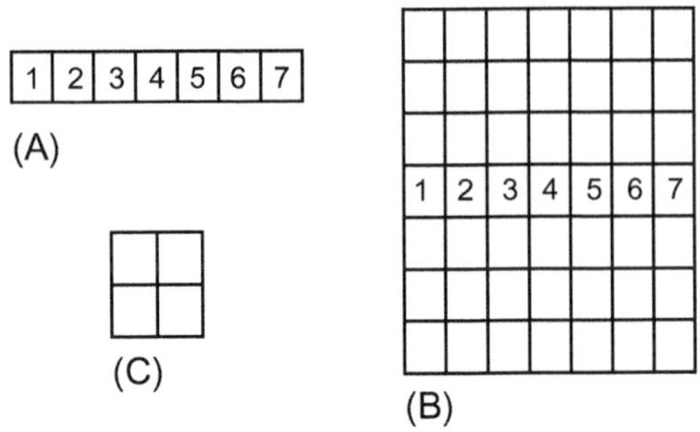

Fig. 2.5 Specimen geographic landscapes. (A) Linear ($n=7$) with its areal units labeled 1–7. (B) Planar 7-by-7 ($n=7\times 7$), with a linear subregion. (C) Planar 2-by-2 ($n=4$).

mathematical link between SA and eigenfunctions in the spectral domain (Section 2.1; Bartlett, 1975). For a one-dimensional geographic landscape (e.g., Fig. 2.5A), given a stationary process $\{X_t, t=1, 2, \ldots, T\}$ for a unidirectional dependency structure (this situation parallels a time series data structure) of the same form as the simultaneous autoregressive (SAR) model specification,

$$X_t = \rho X_{t-1} + \xi_t,$$

where ρ is the autocorrelation parameter, $|\rho|<1$, t indexes location in the linear landscape, and ξ_t is an independent and identically distributed (IID) random error term. Autocovariance functions, $\gamma(\tau)$, $\tau=0, 1, \ldots$, where τ denotes the number of lags, are established by repeatedly substituting the right-hand expression of an autoregressive equation like the preceding one into it and then applying the calculus of expectations to resulting infinite series forms of covariations. These infinite series are functions of linear combinations of ξ_t^2 together with powers of the SA parameter ρ, and hence yield σ^2 as well as a denominator term containing ρ. The corresponding autocovariance (i.e., covariation in an autocorrelated mathematical space) function is defined as follows:

$$\gamma_X(\tau) = \frac{\rho^\tau \sigma^2}{1-\rho^2},$$

with τ being the number of lags (Griffith, 1988, p. 111).

This preceding autocovariance function can be represented in the frequency domain with the Fourier transform.

$$f(\theta) = \frac{1}{2\pi}\sum_{\tau=-\infty}^{\infty} e^{-i\tau\theta}\gamma_X(\tau),$$

where $|\theta| < \pi$ is the frequency, and i is the imaginary part of a complex number. Because $e^{i\tau\theta} = COS(\theta) + iSIN(\theta)$, $f(\theta)$ can be rewritten as follows[5]:

$$f(\theta) = \frac{1}{2\pi}\left[\gamma_X(0) + 2\sum_{\tau=1}^{\infty}\gamma_X(\tau)COS(\theta\tau)\right].$$

Substituting $\gamma_X(\tau)$ in $f(\theta)$ yields

$$f(\theta) = \frac{\sigma^2}{2\pi}\frac{1}{[1 + \rho^2 - 2\rho COS(\theta)]}. \quad (2.5)$$

This representation of the autocovariance function in the frequency domain is referred to as the spectral density function or the spectrum of the random variable (RV) $\{X_t\}$. The sum[6] of the spectrum over all frequencies gives the variance of $\{X_t\}$,

$$VAR(X_t) = \int_0^{\pi} \frac{\sigma^2}{2\pi[1 + \rho^2 - 2\rho COS(\theta)]} d\theta = \frac{\sigma^2}{2(1 - \rho^2)}, \ |\rho| < 1, \quad (2.6)$$

whereas the covariance at lag τ is given by

$$COV(X_t X_{t-1}) = \int_0^{\pi} \frac{\sigma^2 COS(\tau\theta)}{2\pi[1 + \rho^2 - 2\rho COS(\theta)]} d\theta = \frac{\rho^{\tau}\sigma^2}{2(1 - \rho^2)}, \ |\rho| < 1. \quad (2.7)$$

$\tau = 0$ yields Eq. (2.6). In other words, representing the autocovariance function in the frequency domain allows a decomposition of the variance into frequency components that reveals their relative importance. The common effect of variance inflation routinely mentioned as a primary result of SA is apparent here: its arithmetic source is the term $-\rho^2$ appearing in the denominator of this expression. When no SA is present, $\rho = 0$ and $f(\theta) = \sigma^2/2\pi$, which is the spectral density for white noise. Accordingly, the variance in this independent observations case reduces to:

[5] The COS function appearing throughout this discussion is not surprising because COS appears in the mathematics (e.g., algebra, arithmetic, and calculus) of traditional correlation calculations. The Pearson product moment correlation coefficient equals $COS(\theta)$, where θ is the angle of the trend line based upon the z-scores (i.e., subtract the mean, and then divide this difference by the standard deviation) of two attribute RVs X and Y.

[6] Integration is summation of a continuous RV, whereas sigma is summation of a discrete RV.

$$\mathrm{VAR}(X_t) = \int_0^{\pi} \frac{\sigma^2}{2\pi} d\theta = \frac{\sigma^2}{2},$$

which is consistent with results from the classical central limit theorem of statistics.

Furnishing asymptotic details for Eqs. (2.5) and (2.6), as shown by Berman and Plemmons (1994), the eigenvalues of a binary SWM for linear lattices can be defined as

$$\lambda_i = 2\,\mathrm{COS}\left(\frac{i\pi}{n+1}\right).$$

For the spectral density function defined by Eq. (2.5),

$$\lim_{n\to\infty} \frac{i}{n+1} = \theta,$$

$$\lambda_i = 2\mathrm{COS}(\theta), \text{ and}$$

$$f(\theta) = \frac{\sigma^2}{2\pi(1+\rho^2-\rho\lambda)}.$$

In other words, the spectral density function contains the eigenvalues of its corresponding SWM, which reveals that the autocovariance function for the RV $\{X_t\}$ is a function of the eigenvalues of the SWM.

To illustrate the relationship between the lagged spatial correlations with the second-order spatial covariance structure matrix $(\mathbf{I}-\rho\mathbf{C})^{-2}$—the SAR model specification—consider the bidirectional and one-dimensional geographic landscape with a symmetric dependency structure; that is,

$$X_i = \rho(X_{i-1} + X_{i+1}) + \xi_i,\ |\rho| < \frac{1}{2}. \tag{2.8}$$

Here the autovariances at lags 0 and τ are

$$\gamma(0) = E(X_i X_i) = \frac{\sigma^2}{(1-4\rho^2)^{3/2}}, \text{ and}$$

$$\gamma(\tau) = E(X_i X_{i-\tau})$$
$$= \sigma^2 \frac{\mathrm{SIN}[\tau\pi]\ \mathrm{Hypergeometric\ PFQ}\left[\left\{\frac{1}{2}, 1, 2\right\}, \{1-\tau, 1+\tau\} 1-\tau, \frac{4\rho}{1+2\rho}\right]}{\pi\tau(1+2\rho)^2},$$

$$0 < \rho < 1/2$$

where the HypergeometricPFQ function involves Bessel functions[7] (Abramowitz & Stegun, 1972); a second-order specification coupled with bidirectional direct and indirect correlation effects creates a much more complex covariance structure. For this second-order specification, the spectral density function is

$$f(\theta) = \frac{\sigma^2}{2\pi[1 - 2\rho \mathrm{COS}(\theta)]^2},$$

(Griffith, 1988, p. 114) with variance

$$\mathrm{VAR}(X_t) = \int_0^{\pi} \frac{\sigma^2}{2\pi[1 - 2\rho \mathrm{COS}(\theta)]^2} d\theta = \frac{\sigma^2}{2(1 - 4\rho^2)^{3/2}}, \ |\rho| < 1/2. \quad (2.9)$$

Again, if $\rho = 0$, then this variance term reduces to that for white noise, namely, $\sigma^2/2$, and both the exponent 3/2 and $-4\rho^2$ create variance inflation. Meanwhile, the covariance at lag τ is given by

$$\mathrm{COV}(X_t X_{t-1}) = \int_0^{\pi} \frac{\sigma^2 \mathrm{COS}(\tau\theta)}{2\pi[1 - 2\rho \mathrm{COS}(\theta)]^2} d\theta, \ |\rho| < 1. \quad (2.10)$$

This integration yields the preceding HypergeometricPFQ function expression for $\gamma(\tau)$.

The Pearson product moment correlation is defined as the ratio of the covariation between two RVs, X and Y, and the product of their respective standard deviations. Similarly, autocorrelation at lag τ is the ratio between the covariation of RV X and a lagged RV version of X, $\gamma(\tau)$, and the product of the standard deviation of RV X times itself, or the variance of RV X, $\gamma(0)$:

$$\rho(\tau) = \frac{\gamma(\tau)}{\gamma(0)}.$$

In the spectral domain, this ratio is equivalent to the ratio between the integrals of the corresponding spectral density functions:

$$\rho(\tau) = \frac{\int_0^{\pi} g[\mathrm{COS}(\tau\theta)] f(\theta) d\theta}{\int_0^{\pi} f(\theta) d\theta},$$

where g[COS($\tau\theta$)] is simply COS($\tau\theta$) in one dimension and becomes COS($\tau\theta$)COS($\eta\varphi$) in two dimensions, where η denotes the number of lags

[7] This expression reduces to $\frac{2\rho\sigma^2}{(1-4\rho^2)^{3/2}}$ for $\tau = 1$.

and φ denotes the frequency in the second dimension for a regular square tessellation forming a complete rectangular region and a rook's adjacency definition of contiguity. Eqs. (2.8)–(2.10) pertain to SAR models and a linear landscape. The simpler conditional autoregressive (CAR) model specification, which involves first-order spatial dependency, has the following autocorrelation function:

$$\rho(\tau) = \frac{\int_0^{\pi} \frac{\sigma^2 \mathrm{COS}(\tau\theta)}{2\pi[1-2\rho \mathrm{COS}(\theta)]} d\theta}{\int_0^{\pi} \frac{\sigma^2}{2\pi[1-2\rho \mathrm{COS}(\theta)]} d\theta},$$

whose solution again involves the HypergeometricPFQ function. Table 2.3 lists the specific spatial autocorrelation formulae at lags $\tau=0$ through $\tau=6$. The pattern of distance decay is equivalent to the one generated through the matrix $(\mathbf{I}-\rho\mathbf{C})^{-1}/v_{kk}$, referred to as the spatial correlation matrix; v_{kk} is the center cell of the matrix $(\mathbf{I}-\rho\mathbf{C})$, which can be expanded as the following infinite series:

$$\mathbf{I} + \rho\mathbf{C} + \rho^2\mathbf{C}^2 + \rho^3\mathbf{C}^3 + \ldots,$$

furnishing a different expression of distance decay in terms of movements across boundaries of a surface partitioning. Table 2.4 details the spatial correlation matrix $(\mathbf{I}-0.25\mathbf{C})^{-1}/v_{44}$, where v_{44} is the central cell value of the 7-by-7 matrix $(\mathbf{I}-0.25\mathbf{C})^{-1}$. The diagonal entries are not 1, except for the areal unit #4 correlation. These deviations from 1 are attributable to edge effects. These latter correlations converge on their spectral density-based counterparts as n goes to infinity.

Table 2.5 reports the spatial correlation matrix, say $\mathbf{V}$, constructed from the spectral-based lag spatial correlations (Table 2.3). Here all diagonal entries are 1. For comparison purposes, Table 2.6 reports the equivalent to $(\mathbf{I}-0.25\mathbf{C})^{-1}$ based upon the spectral density function correlations, confirming a relationship between the two conceptualizations. The two end areal units in the specimen linear landscape have values of 0.933013 rather than 1 because of edge effects. This example demonstrates that matrix inversion uncovers edge effects and that these distortions are more severe for SWM descriptions of SA.

Meanwhile, for a second-order specification (i.e., the SAR model), which captures a wider autocorrelation range and a more intense SA field, Eqs. (2.9) and (2.10) relate to the SWM specification $(\mathbf{I}-\rho\mathbf{C})^2$. Again, the second-order perspective pertains to the power of 2 in both of these

Table 2.3 Lag spatial correlation based upon the first-order spectral density function for a linear geographic landscape

Lag	Spectral density-based correlation equation: First-order spatial dependency	$\rho = 1/4$	Central unit for $(I - 0.25C)^{-1}$
0	1	1	1
1	$\left[1-\sqrt{1-4\rho^2}\right]/(2\rho)$	0.267949	0.267857
2	$\left[1-2\rho^2-\sqrt{1-4\rho^2}\right]/(2\rho^2)$	0.071797	0.071747
3	$\left[1-3\rho^2-(1-\rho^2)\sqrt{1-4\rho^2}\right]/(2\rho^3)$	0.019238	0.017857
4	$\left[1-4\rho^2+2\rho^4-(1-2\rho^2)\sqrt{1-4\rho^2}\right]/(2\rho^4)$	0.005155	
5	$\left[1-5\rho^2+5\rho^4-(1-3\rho^2+\rho^4)\sqrt{1-4\rho^2}\right]/(2\rho^5)$	0.001381	
6	$\left[1-6\rho^2+9\rho^4-2\rho^6-(1-4\rho^2+3\rho^4)\sqrt{1-4\rho^2}\right]/(2\rho^6)$	0.000370	

Table 2.4 $(\mathbf{I}-\rho\mathbf{C})^{-1}/v_{44}$, where $v_{44}=1.15464$ is the central cell value of 7-by-7 matrix $(\mathbf{I}-\rho\mathbf{C})^{-1}$, $\rho=1/4$ (maximum value$=0.5$)

Areal unit	1	2	3	4	5	6	7
1	0.928253	0.248724	0.066645	0.01786	0.004783	0.001276	0.000319
2	0.248724	0.994898	0.266582	0.07143	0.019133	0.005102	0.001276
3	0.066645	0.266582	0.999681	0.26786	0.071747	0.019133	0.004783
4	0.017857	0.071429	0.267857	1.00000	0.267857	0.071429	0.017857
5	0.004783	0.019133	0.071747	0.26786	0.999681	0.266582	0.066645
6	0.001276	0.005102	0.019133	0.07143	0.266582	0.994898	0.248724
7	0.000319	0.001276	0.004783	0.01786	0.066645	0.248724	0.928253

Table 2.5 Matrix **V** based upon spectral density function correlations, $\rho = 1/4$ (maximum value $= 0.5$)

Areal unit	1	2	3	4	5	6	7
1	1.00000	0.26795	0.07180	0.01924	0.00515	0.00138	0.00037
2	0.26795	1.00000	0.26795	0.07180	0.01924	0.00515	0.00138
3	0.07180	0.26795	1.00000	0.26795	0.07180	0.01924	0.00515
4	0.01924	0.07180	0.26795	1.00000	0.26795	0.07180	0.01924
5	0.00515	0.01924	0.07180	0.26795	1.00000	0.26795	0.07180
6	0.00138	0.00515	0.01924	0.07180	0.26795	1.00000	0.26795
7	0.00037	0.00138	0.00515	0.01924	0.07180	0.26795	1.00000

Table 2.6 Matrix $\mathbf{V}^{-1}/w_{44}$, where $w_{44}=1.15470$ is the central cell value of matrix $\mathbf{V}^{-1}$, $\rho=1/4$ (maximum value $=0.5$)

Areal unit	1	2	3	4	5	6	7
1	0.933013	−0.25000	−0.00000	0.00000	−0.00000	0.00000	−0.000000
2	−0.250000	1.00000	−0.25000	−0.00000	0.00000	−0.00000	0.000000
3	−0.000000	−0.25000	1.00000	−0.25000	−0.00000	0.00000	−0.000000
4	0.000000	−0.00000	−0.25000	1.00000	−0.25000	−0.00000	0.000000
5	−0.000000	0.00000	−0.00000	−0.25000	1.00000	−0.25000	−0.000000
6	0.000000	−0.00000	0.00000	−0.00000	−0.25000	1.00000	−0.250000
7	−0.000000	0.00000	−0.00000	0.00000	−0.00000	−0.25000	0.933013

Table 2.7 Lag spatial correlations for areal unit #4 in the specimen linear landscape, $\rho = 1/4$ (maximum value = 0.5)

Lag	Spectral density	Central unit for $(\mathbf{I} - 0.25\mathbf{C})^{-2}$	Central unit for $[(\mathbf{I} - 0.25\mathbf{W}^T)(\mathbf{I} - 0.25\mathbf{W})]^{-1}$
0	1	1	1
1	0.5	0.499448	0.50111
2	0.196152	0.193923	0.20141
3	0.069219	0.061879	0.11423

equations and the SWM matrix expression. Table 2.7 presents the lag spatial correlations for the central areal unit, confirming that these correlations are larger.

Insights also can be gleaned from the row-standardized SWM **W**. Griffith (1980) notes that its eigenvalues for a linear geographic landscape are as follows:

$$\lambda_i = \mathrm{COS}[i\pi/(n-1)],\ i = 0, 1, 2, \ldots, n-1$$

(also see Berman & Plemmons, 1994). Accordingly, for equations such as (2.7) and (2.10), the SA parameter becomes 2ρ, rather than 0.25, and the example SA parameter becomes 0.5. Consequently, the spectral density-based correlations in Table 2.7 do not change. However, the spatial lag correlations based upon matrix **W** exhibit even more severe edge effects than those based upon matrix **C**. This severity is due, in part, to the small sample size (i.e., $n = 7$) involved here.[8]

The eigenvectors of SWM **C** relate to SA. Transforming matrix **C** to matrix $(\mathbf{I} - \rho\mathbf{C})$ does not change this relationship: the eigenvalues of matrix **C** become $1 - \rho\lambda_i$, $i = 1, 2, \ldots, n$; and, the eigenvectors of matrices **C** and $(\mathbf{I} - \rho\mathbf{C})$ are identical. In addition, the spectral density for spatial lag τ is an entry in the n-by-n matrix **V**, which asymptotically equals $(\mathbf{I} - \rho\mathbf{C})^{-1}$. Consequently, theoretically the eigenvectors of these two matrices are identical. Fig. 2.6 corroborates this equivalency for the case of $n = 7$ and $\rho = 0.25$; because of symmetry in the linear geographic landscape, each eigenvector has three pairs of values, resulting in only four distinct values in a vector. The sets of corresponding values almost perfectly align in a straight line (i.e., collinear), with slight deviations attributable to edge effects. Furthermore, because eigenvectors are unique to a multiplicative factor of -1, Fig. 2.6A exhibits a negative relationship, whereas Fig. 2.6B exhibits a positive relationship between the respective pairs of eigenvectors.

[8] See Tiefelsdorf, Griffith, and Boots (1999) for a treatment of this situation for two-dimensional geographic landscapes.

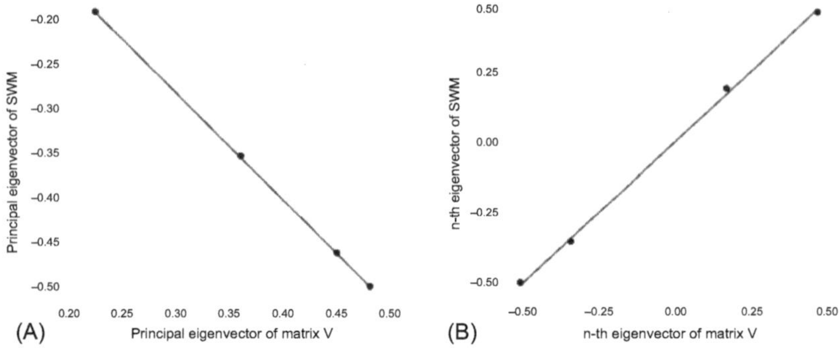

Fig. 2.6 Scatterplots of corresponding eigenvectors from SWM **C** and matrix **V** constructed with the spectral density function-based Eq. (2.8) for the specimen linear landscape. (A) Principal eigenvalues. (B) Eigenvectors $\mathbf{E}_7$.

Geostatistics works directly with spectral density results, whereas spatial autoregression works directly with spatial linear operators like $(\mathbf{I}-\rho\mathbf{C})$ constructed with SWMs. One minus the correlations rendered by a spectral density function are the quantities treated in geostatistics. Accordingly, semivariance, in terms of lagged spatial correlation, is

$$1-\int_0^{\pi}\frac{\mathrm{COS}(k\theta)}{[1-2\rho\mathrm{COS}(\theta)]^{\tau}}d\theta \bigg/ \int_0^{\pi}\frac{1}{[1-2\rho\mathrm{COS}(\theta)]^{\tau}}d\theta.$$

For $\rho=0.25$ and the CAR model specification (i.e., $\tau=1$ in the spectral density equation), spatial correlation goes to zero at lag 19, and the semivariogram trend line description is the following estimated exponential model, which has a nugget of zero:

$$1-e^{-\mathrm{distance}/0.75933},\ \text{pseudo-R}^2=1.$$

Meanwhile, for $\rho=0.25$ and the SAR model specification (i.e., $\tau=2$ in the spectral density equation), spatial correlation goes to zero at lag 21, and the semivariogram trend line description is the following estimated Matérn-type Bessel function (B) model, which also has a nugget of zero:

$$1-B\left(\mathrm{distance}^{1.20871}/0.74978\right),\ \text{pseudo-R}^2=1.$$

Fig. 2.7 portrays the corresponding semivariogram plots with their respective estimated model prediction values overlaid on them.

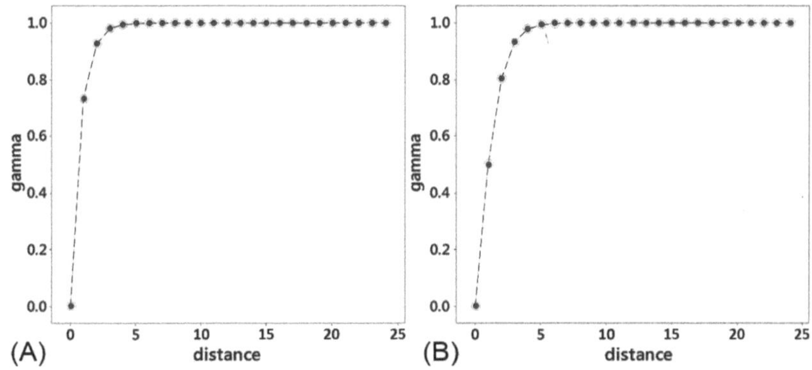

Fig. 2.7 Semivariogram plot trend line values superimposed on adjusted spectral density function values for a linear geographic landscape. (A) The CAR results for $\rho=0.25$. (B) The SAR results for $\rho=0.25$.

2.3 The spectral analysis of two-dimensional data

Griffith and Layne (1999, pp. 142–52) outline and discuss the spectral density function conceptualization for planar geographic landscapes, noting that the spectral density-based spatial (h,k)-lag correlation function becomes, for a regular square tessellation and the rook adjacency definition,

$$\int_0^{\pi}\int_0^{\pi}\frac{\cos(\tau\theta)\cos(\eta\varphi)}{\{1-\rho 2[\cos(\theta)+\cos(\varphi)]\}^{\kappa}}d\theta d\varphi \Big/ \int_0^{\pi}\int_0^{\pi}\frac{1}{\{1-\rho 2[\cos(\theta)+\cos(\varphi)]\}^{\kappa}}d\theta d\varphi,$$

where $\kappa=1$ is for a first-order specification (which links to a CAR model and an exponential semivariogram model), and $\kappa=2$ is for a second-order specification (which links to a SAR model and a K-Bessel function semivariogram model). As before, the denominator contains the eigenvalues of SWM **C**. If $\tau=\eta=0$, then this expression equals 1, as it should. Table 2.8 reports comparable calculations from spectral density and 7^2-by-$7^2=49$-by-49 matrix inversion computations.

Fig. 2.8 portrays the extreme eigenvectors for the 7-by-7 regular square tessellation geographic landscape (see Fig. 2.5B). MC$=0.95141$ (a value that is <1) for the maximum PSA cases appearing in Fig. 2.8A and B: these map patterns (which are the second and third eigenvectors of SWM **C**) are gradients across the geographic landscape, one trend being from the northeast to the southwest, and the other being from the northwest to the southeast (they

Table 2.8 (h,0)-lag spatial correlations for areal unit #4 in linear landscape embedded in the planar geographic landscape, $\rho = 1/8$ (maximum value $= 0.25$)

Lag	Spectral density, $\kappa = 1$	Central unit for $(I - 0.125C)^{-1}$	Spectral density, $\kappa = 2$	Central unit for $(I - 0.125C)^{-2}$
0,0	1	1	1	1
1,0	0.136383	0.136381	0.276872	0.276856
2,0	0.019223	0.019213	0.059977	0.059898
3,0	0.002791	0.002728	0.011888	0.011467

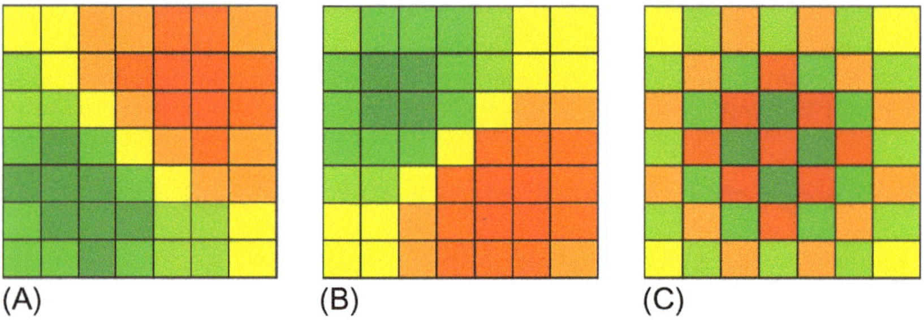

Fig. 2.8 The extreme MC eigenvectors for a 7-by-7 regular square tessellation surface partitioning. (A) MC$=$0.95141. (B) MC$=$0.95141. (C) MC$=-$1.07786.

are orthogonal, one being a 90 degrees rotation of the other).[9] Fig. 2.8C depicts the maximum NSA possible (MC$=-1.07786$). It also better highlights the presence of edge effects, exhibiting a central set of stark contrasts surrounded by intermediate values. These edge effects appear in Fig. 2.8A and B, too: the *dark red* and *green pixels* are not in the extreme corners of these surface partitionings.

Returning to a geostatistics perspective, analyzing the semivariance, in terms of two-dimensional lagged spatial correlation for $\rho = 0.125$ and the CAR model specification (i.e., $\kappa = 1$ in the spectral density equation), spatial correlation goes to zero at lag 13, and the semivariogram trend line description is the following estimated exponential model, which has a nugget of zero:

$$1 - e^{-\text{distance}/0.49597},\ \text{pseudo-}R^2 = 0.9995.$$

Meanwhile, for $\rho = 0.125$ and the SAR model specification (i.e., $\kappa = 2$ in the spectral density equation), spatial correlation goes to zero at lag 15, and the

[9] These eigenvectors are not the correct theoretical ones, which exhibit east–west and north–south map patterns.

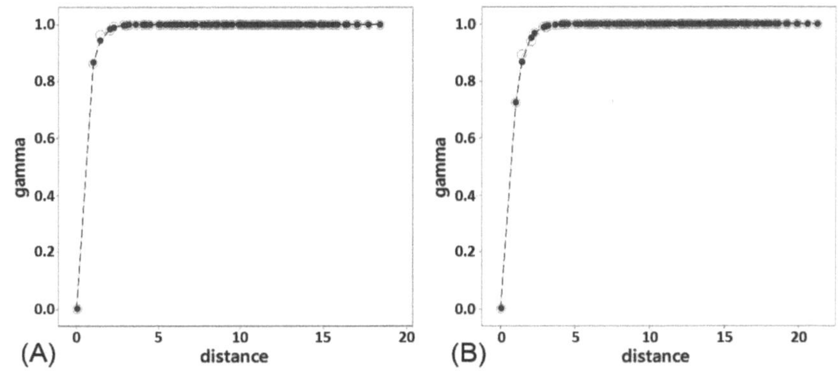

Fig. 2.9 Semivariogram plot trend line values superimposed on adjusted spectral density function values for a two-dimensional geographic landscape. (A) The CAR results for $\rho=0.125$. (B) The SAR results for $\rho=0.125$.

semivariogram trend line description is the following estimated Matérn-type Bessel function model, which also has a nugget of zero:

$$1-\mathrm{B}\left(\text{distance}^{0.52078}/0.58926\right),\ \text{pseudo-R}^2=0.9990.$$

Fig. 2.9 portrays the corresponding semivariogram plots with their respective estimated model prediction values overlaid on them. The slight predicted-versus-observed discrepancies here are for the lag 3 value pairs.

2.4 The spectral analysis of three-dimensional data

In many cases, the spectral analysis of three-dimensional georeferenced data involves a sequence of maps, one for each point in a specified time series, rather than supplementing planar surfaces with elevation. Griffith and Heuvelink (2012) extend the preceding spectral analysis conceptualizations to this situation. Now the spectral density-based space–time (τ,η,ν)-lag correlation function becomes, for a regular square tessellation and the rook adjacency definition, and uniformly spaced points in time,

$$\frac{\displaystyle\int_0^{\pi}\int_0^{\pi}\int_0^{\pi}\frac{\mathrm{COS}(\tau\theta)\mathrm{COS}(\eta\varphi)\mathrm{COS}(\nu t)}{[1-\mathrm{COS}(t)\{\rho_s[\mathrm{COS}(\theta)+\mathrm{COS}(\varphi)]+\rho_T\}]^{\kappa}}d\theta d\varphi dt}{\displaystyle\int_0^{\pi}\int_0^{\pi}\int_0^{\pi}\frac{1}{[1-\mathrm{COS}(t)\{\rho_s[\mathrm{COS}(\theta)+\mathrm{COS}(\varphi)]+\rho_T\}]^{\kappa}}d\theta d\varphi dt},\ \kappa=1,2,$$

where t denotes the time argument, ρ_s denotes the SA parameter, and ρ_T denotes the temporal autocorrelation parameter. This specification represents a contemporaneous space–time process, which is additive, whose matrix representation is given by

$$\mathbf{C} = \mathbf{I}_T \otimes \mathbf{I}_s - \rho_s \mathbf{C}_T \otimes \mathbf{C}_s - \rho_T \mathbf{C}_T \otimes \mathbf{I}_s,$$

where $\otimes$ denotes the Kronecker product mathematical matrix operation, $\mathbf{C}_s$ denotes the SWM, $\mathbf{C}_T$ denotes the time-series connectivity matrix, $\mathbf{I}_T$ denotes the T-by-T identify matrix, $\mathbf{I}_s$ denotes the n-by-n identity matrix, and $1 - \mathrm{COS}(t)\{\rho_s[\mathrm{COS}(u) + \mathrm{COS}(v)] + \rho_T\}$ are the limiting eigenvalues of the space–time connectivity matrix $\mathbf{C}$.

An alternative specification is multiplicative and hence describes a space–time lagged process; its matrix representation is given by

$$\mathbf{C} = \mathbf{I}_T \otimes \mathbf{I}_s - \rho_s \mathbf{I}_T \otimes \mathbf{C}_s - \rho_T \mathbf{C}_T \otimes \mathbf{I}_s,$$

and its spectral density-based (τ,η,ν)-lag correlations are given by

$$\frac{\displaystyle\int_0^\pi \int_0^\pi \int_0^\pi \frac{\mathrm{COS}(\tau\theta)\mathrm{COS}(\eta\varphi)\mathrm{COS}(\nu t)}{\{1 - \rho_s[\mathrm{COS}(\theta) + \mathrm{COS}(\varphi)] - \rho_T\,\mathrm{COS}(t)\}^\kappa} d\theta d\varphi dt}{\displaystyle\int_0^\pi \int_0^\pi \int_0^\pi \frac{1}{\{1 - \rho_s[\mathrm{COS}(\theta) + \mathrm{COS}(\varphi)] - \rho_T\,\mathrm{COS}(t)\}^\kappa} d\theta d\varphi dt}, \quad \kappa = 1, 2.$$

For a regular square lattice forming a complete P-by-Q rectangular region,

$$\mathbf{C}_s = \mathbf{C}_P \otimes \mathbf{I}_Q + \mathbf{C}_Q \otimes \mathbf{I}_P,$$

where $\mathbf{C}_P$ and $\mathbf{C}_Q$, respectively, are SWMs for a P length and a Q length linear landscape, and $\mathbf{I}_P$ and $\mathbf{I}_Q$, respectively, are P-by-P and Q-by-Q identity matrices.

2.5 Summary

This chapter reviews articulations among SWMs, eigenfunctions, and spectral functions, all three of which relate to SA. In doing so, it also links them to geostatistics. The eigenvalues of a SWM index the nature and degree of SA in the eigenvectors of a modified SWM and also appear in the complex fraction spectral density functions used to calculate lagged spatial correlations. The cells of standardized inverse spatial covariance structures, illustrated here with the popular first- and second-order ones, contain spectral density function results. These notions interlace with concepts for PCA. Although this chapter focuses on the MC index of SA, similar results may be established for both the Geary ratio (GR) and the join count statistics that are applicable to nominal measurement scale data. The linear geographic landscape furnishes many relatively simple illustrations of the connections of interest here. The two-dimensional geographic landscape furnishes more relevant, albeit more complicated, contexts and highlights map pattern visualizations, one of the most important topics of this chapter.

Appendix 2.A The spectral decomposition of a SWM

Consider the geographic landscape in Fig. 2.5C. Its rook adjacency SWM **C** is as follows:

$$\begin{bmatrix} 0 & 1 & 1 & 0 \\ 1 & 0 & 0 & 1 \\ 1 & 0 & 0 & 1 \\ 0 & 1 & 1 & 0 \end{bmatrix}.$$

The Perron–Frobenius theorem states that the principal eigenvalue is contained in the interval defined by the largest and smallest row sums; therefore here $\lambda_1 = 2$. For each pair of rows or columns that is identical, an eigenvalue equals zero; therefore because the first and fourth rows/columns are identical, and the second and third rows/columns are identical, two eigenvalues equal zero. Finally the trace of this matrix equals the sum of its four eigenvalues; therefore $2+0+0+\lambda=0$, and hence an eigenvalue equals -2.

Eq. (2.1) for this SWM is $\lambda^2(\lambda^2-4)=0$. The first λ^2 term is for the two roots of zero, whereas the second term factors into $(\lambda+2)(\lambda-2)$, which is for the two roots ± 2. Ord (1975) also states that the eigenvalues for this particular type of geographic surface partitioning and SWM are given by $\lambda = 2\left[\mathrm{COS}\left(\frac{h\pi}{2+1}\right)+\mathrm{COS}\left(\frac{k\pi}{2+1}\right)\right]$, h=1, 2 and k=1, 2. This equation yields $2(0.5+0.5)=2$; $2(0.5-0.5)=0$; $2(-0.5+0.5)=0$; and, $2(-0.5-0.5)=-2$.

Griffith (2000, p. 98) proves that the solution to Eq. (2.2) for this particular type of geographic surface partitioning and SWM are the eigenvectors given by

$$\frac{2}{\sqrt{(2+1)(2+1)}}\left[\mathrm{SIN}\left(\frac{h\pi}{2+1}\right)\times\mathrm{SIN}\left(\frac{k\pi}{2+1}\right)\right]$$

This expression produces the 4-by-4 eigenvector matrix

$$\begin{bmatrix} 0.5 & 0.5 & 0.5 & 0.5 \\ 0.5 & -0.5 & 0.5 & -0.5 \\ 0.5 & 0.5 & -0.5 & -0.5 \\ 0.5 & -0.5 & -0.5 & 0.5 \end{bmatrix}$$

Consequently, the SWM eigendecomposition (see Section 2.1.3) is the following:

$$\begin{bmatrix} 0.5 & 0.5 & 0.5 & 0.5 \\ 0.5 & -0.5 & 0.5 & -0.5 \\ 0.5 & 0.5 & -0.5 & -0.5 \\ 0.5 & -0.5 & -0.5 & 0.5 \end{bmatrix} \bullet \begin{bmatrix} 2 & 0 & 0 & 0 \\ 0 & 0 & 0 & 0 \\ 0 & 0 & 0 & 0 \\ 0 & 0 & 0 & -2 \end{bmatrix} \bullet \begin{bmatrix} 0.5 & 0.5 & 0.5 & 0.5 \\ 0.5 & -0.5 & 0.5 & -0.5 \\ 0.5 & 0.5 & -0.5 & -0.5 \\ 0.5 & -0.5 & -0.5 & 0.5 \end{bmatrix} =$$

$$\begin{bmatrix} 1 & 0 & 0 & -1 \\ 1 & 0 & 0 & 1 \\ 1 & 0 & 0 & 1 \\ 1 & 0 & 0 & -1 \end{bmatrix} \bullet \begin{bmatrix} 0.5 & 0.5 & 0.5 & 0.5 \\ 0.5 & -0.5 & 0.5 & -0.5 \\ 0.5 & 0.5 & -0.5 & -0.5 \\ 0.5 & -0.5 & -0.5 & 0.5 \end{bmatrix} = \begin{bmatrix} 0 & 1 & 1 & 0 \\ 1 & 0 & 0 & 1 \\ 1 & 0 & 0 & 1 \\ 0 & 1 & 1 & 0 \end{bmatrix}$$

The general form of these algebraic mechanics is $\mathbf{E\Lambda E}^{\mathrm{T}} = \mathbf{C}$.

For matrix $\mathbf{C} + \mathbf{I}$, the SWM matrix analyzed in this appendix becomes

$$\begin{bmatrix} 1 & 1 & 1 & 0 \\ 1 & 1 & 0 & 1 \\ 1 & 0 & 1 & 1 \\ 0 & 1 & 1 & 1 \end{bmatrix}$$

The Perron–Frobenius theorem yields $\lambda_1 = 3$. The matrix determinant yields the polynomial equation

$$(\lambda + 1)(\lambda - 1)^2(\lambda - 3) = 0,$$

whose solution is the set of roots $\{-1, 1, 1, 3\}$; this set of roots is exactly the set $\{-2+1, 0+1, 0+1, 2+1\}$. In addition, the roots sum to four, the trace of this augmented SWM. Meanwhile,

$$\begin{bmatrix} 0.5 & 0.5 & 0.5 & 0.5 \\ 0.5 & -0.5 & 0.5 & -0.5 \\ 0.5 & 0.5 & -0.5 & -0.5 \\ 0.5 & -0.5 & -0.5 & 0.5 \end{bmatrix} \bullet \begin{bmatrix} 3 & 0 & 0 & 0 \\ 0 & 1 & 0 & 0 \\ 0 & 0 & 1 & 0 \\ 0 & 0 & 0 & -1 \end{bmatrix} \bullet \begin{bmatrix} 0.5 & 0.5 & 0.5 & 0.5 \\ 0.5 & -0.5 & 0.5 & -0.5 \\ 0.5 & 0.5 & -0.5 & -0.5 \\ 0.5 & -0.5 & -0.5 & 0.5 \end{bmatrix} =$$

$$\begin{bmatrix} 1.5 & 0.5 & 0.5 & 0.5 \\ 1.5 & -0.5 & 0.5 & -0.5 \\ 1.5 & 0.5 & -0.5 & -0.5 \\ 1.5 & -0.5 & -0.5 & 0.5 \end{bmatrix} \bullet \begin{bmatrix} 0.5 & 0.5 & 0.5 & 0.5 \\ 0.5 & -0.5 & 0.5 & -0.5 \\ 0.5 & 0.5 & -0.5 & -0.5 \\ 0.5 & -0.5 & -0.5 & 0.5 \end{bmatrix} = \begin{bmatrix} 1 & 1 & 1 & 0 \\ 1 & 1 & 0 & 1 \\ 1 & 0 & 1 & 1 \\ 0 & 1 & 1 & 1 \end{bmatrix}$$

In other words, $\mathbf{E}(\mathbf{\Lambda} + \mathbf{I})\mathbf{E}^{\mathrm{T}} = \mathbf{C} + \mathbf{I}$, where eigenvector matrix $\mathbf{E}$ is the same as for matrix $\mathbf{C}$ alone.

References

Abramowitz, M., & Stegun, I. (Eds.), (1972). Bessel functions of integer order Chapter 9. In *Handbook of mathematical functions with formulas, graphs, and mathematical tables* (pp. 355–389). New York: Dover.

Bartlett, M. (1975). *The statistical analysis of spatial pattern.* London: Chapman and Hall.
Berman, A., & Plemmons, R. (1994). *Nonnegative matrices in the mathematical sciences* [corrected republication of the1979 edition]. Philadelphia: SIAM.
Griffith, D. (1980). Towards a theory of spatial statistics. *Geographical Analysis, 12*, 325–339.
Griffith, D. (1984). Measuring the arrangement property of a system of areal units generated by partitioning a planar surface. In G. Bahrenberg, M. Fischer, & P. Nijkamp (Eds.), *Recent developments in spatial analysis: Methodology, measurement, models* (pp. 191–200). Aldershot, UK: Gower.
Griffith, D. (1988). *Advanced spatial statistics.* Dordrecht: Martinus Nijhoff.
Griffith, D. (1996). Spatial autocorrelation and eigenfunctions of the geographic weights matrix accompanying geo-referenced data. *Canadian Geographer, 40*, 351–367.
Griffith, D. (2000). Eigenfunction properties and approximations of selected incidence matrices employed in spatial analyses. *Linear Algebra and Its Applications, 321*, 95–112.
Griffith, D., & Heuvelink, G. (2012). Deriving space–time variograms from space–time autoregressive (STAR) model specifications. Chapter 1 In A. Yeh, J. Shi, Y. Leung, & C. Zhou (Eds.), *Advances in spatial data handling and GIS* (pp. 3–14). Berlin: Springer Verlag.
Griffith, D., & Layne, L. (1999). *A casebook for spatial statistical data analysis.* New York: Oxford University Press.
Ord, K. (1975). Estimation methods for models of spatial interactions. *Journal of the American Statistical Association, 70*, 120–126.
Peria, S. 2007. Dispersion of light through a prism. Wolfram Demonstrations Project, http://demonstrations.wolfram.com/DispersionOfLightThroughAPrism/.
Rayner, J. (1971). *Introduction to spectral analysis.* London: Pion.
Tiefelsdorf, M., & Boots, B. (1995). The exact distribution of Moran's I. *Environment and Planning A*, 27(6), 985–999.
Tiefelsdorf, M., Griffith, D., & Boots, B. (1999). A variance stabilizing coding scheme for spatial link matrices. *Environment and Planning A, 31*, 165–180.

CHAPTER 3

MESF and linear regression

Chapter outline

The notion of linear regression dates back to Galton and 1885 (Gorroochurn, 2015) and was the first variety of regression both receiving rigorous treatment and being employed widely by scientific data analysts. It became the workhorse of classical statistics; virtually all statistical techniques could be rewritten as a regression problem (the general linear model today), although such formulations often were less efficient. Meanwhile, the notion of spatial filtering dates back to Griffith (1978), first appearing in the tradition of Cochrane–Orcutt (1949) time-series filtering; Moran eigenvector spatial filtering (MESF) dates back to Griffith (1996), with its first rigorous linear regression treatment appearing in Griffith (2000a).

In its simplest form, linear regression is a straight line algebraic model specification for describing the correspondence between a scalar response variable denoted by Y_i and one or more covariates denoted by a row vector $\mathbf{X}_i$ for a single observation i; multiple observations result in concatenation of these items, forming a vector $\mathbf{Y}$ and a matrix $\mathbf{X}$. Bivariate or simple linear regression is the case involving only a single covariate; multiple linear regression is the case involving more than one covariate. In either specification, an n-by-1 vector of ones, $\mathbf{1}$, is included as a covariate for the

Spatial Regression Analysis Using Eigenvector Spatial Filtering
https://doi.org/10.1016/B978-0-12-815043-6.00003-3

intercept term, where n is the number of areal units. Meanwhile, in its simplest form, eigenvector spatial filtering is built upon the Moran coefficient (MC) and thus may be referred to as MESF. Its foundation is a linear regression model specification, and its name refers to it filtering spatial autocorrelation (SA) out of regression residuals and then transferring this SA to the regression intercept term. By doing so, the intercept term no longer is a constant, and the residuals are able to mimic independent, uncorrelated ones. The eigenvectors used to do this spatial filtering come from the following modified spatial weights matrix (SWM) appearing in the numerator of the MC:

$$\left(\mathbf{I} - \mathbf{1}\mathbf{1}^{\mathrm{T}}/\mathrm{n}\right)\mathbf{C}\left(\mathbf{I} - \mathbf{1}\mathbf{1}^{\mathrm{T}}/\mathrm{n}\right), \tag{3.1}$$

where $\mathbf{I}$ is the n-by-n identity matrix, and $\mathbf{C}$ is the n-by-n SWM for a given geographic landscape. An eigenvector spatial filter (ESF) is a linear combination of some subset of these n eigenvectors. Coefficients in this linear combination can be calculated with linear regression. Because SA latent in most georeferenced data relates to a number of eigenvectors, K, which is substantially less than n (i.e., $K << n$), the technique required to construct an ESF almost always is multiple linear regression implemented with a stepwise selection of eigenvectors procedure.

3.1 A theoretical foundation for ESFs

The theoretical foundation for MESF contains two components, one derivable from the general spatial autoregressive model specification and the other derivable from the concept of a random effects term.

The spatial autoregressive response (AR) model (known as the spatial lag model in spatial econometrics) specification, an auto-normal model, may be written as follows, using the spatial linear operator $(\mathbf{I} - \rho\mathbf{C})$ and matrix notation:

$$\mathbf{Y} = (\mathbf{I} - \rho\mathbf{C})^{-1}(\mathbf{X}\boldsymbol{\beta}_{\mathbf{X}} + \boldsymbol{\varepsilon}), \tag{3.2}$$

where $\boldsymbol{\beta}_{\mathbf{X}}$ is a (p + 1)-by-1 vector of regression coefficients for p covariates and the intercept term, ρ is the SA parameter, and $\boldsymbol{\varepsilon}$ is an n-by-1 vector of independent and identically distributed (IID) normal random variables (RVs) with mean zero and constant variance σ^2. The standard maximum likelihood estimation of parameters in Eq. (3.2) involves it being rewritten as the following nonlinear regression specification:

$$(\mathbf{I}-\rho\mathbf{C})\mathbf{Y}=(\mathbf{I}-\rho\mathbf{C})(\mathbf{I}-\rho\mathbf{C})^{-1}(\mathbf{X}\boldsymbol{\beta}_{\mathbf{X}}+\boldsymbol{\varepsilon})\Rightarrow\mathbf{Y}=\rho\mathbf{C}\mathbf{Y}+\mathbf{X}\boldsymbol{\beta}_{\mathbf{X}}+\boldsymbol{\varepsilon}. \tag{3.3}$$

The eigenfunction decomposition of the SWM $\mathbf{C}$ is $\mathbf{E}\boldsymbol{\Lambda}\mathbf{E}^{\mathrm{T}}$, where matrix $\mathbf{E}$ is the set of n eigenvectors of SWM $\mathbf{C}$, diagonal matrix $\boldsymbol{\Lambda}$ contains the set of n eigenvalues of SWM $\mathbf{C}$, with the ordering of entries in these two matrices being the same eigenfunctions, and superscript T denotes the matrix transpose operation. Substituting this decomposition of SWM $\mathbf{C}$ into Eq. (3.3) produces.

$$\mathbf{Y}=\rho\mathbf{E}\boldsymbol{\Lambda}\mathbf{E}^{\mathrm{T}}\mathbf{Y}+\mathbf{X}\boldsymbol{\beta}_{\mathbf{X}}+\boldsymbol{\varepsilon},$$

where $\mathbf{E}^{\mathrm{T}}\mathbf{Y}$ is the ordinary least squares (OLS) estimate of regression coefficients when response variable $\mathbf{Y}$ is regressed on eigenvector matrix $\mathbf{E}$. A stepwise selection procedure (e.g., simultaneous forward-backward) eliminates j eigenvectors for which $\mathbf{E}_{\mathrm{j}}^{\mathrm{T}}\mathbf{Y}\approx 0$ (i.e., the SA map patterns for these eigenvectors do not account for any SA in the regression residuals) or for which $\rho\lambda_{\mathrm{j}}\approx 0$ (i.e., the map pattern displays a trivial degree of SA), which in practice tends to be a large majority of the eigenvectors, leaving $\mathrm{K}<<\mathrm{n}$ eigenvectors in the model specification:

$$\mathbf{Y}=\mathbf{E}_{\mathrm{K}}\boldsymbol{\beta}_{\mathbf{E}}+\mathbf{X}\boldsymbol{\beta}_{\mathbf{X}}+\boldsymbol{\xi}, \tag{3.4}$$

where $\boldsymbol{\beta}_{\mathbf{E}}$ is an n-by-K vector of coefficients used to construct the ESF $\mathbf{E}_{\mathrm{K}}\boldsymbol{\beta}_{\mathbf{E}}$ for attribute variable Y, and $\boldsymbol{\xi}$ is an n-by-1 vector of error terms that mimics $\boldsymbol{\varepsilon}$, a set of IID normal RVs with mean zero and constant variance σ^2. Because of degrees of freedom concerns, devising this formulation promotes first establishing a candidate set from which to select eigenvectors with a stepwise procedure. This candidate set can be based upon the nature of SA exhibited by the regression residuals $\mathbf{Y}-\mathbf{X}\mathbf{b}_{\mathbf{X}}$, where the (p + 1)-by-1 vector $\mathbf{b}_{\mathbf{X}}$ is the OLS estimate of $\boldsymbol{\beta}_{\mathbf{X}}$. Chun et al. (2016) argue that it also should be based upon the degree of SA exhibited by a set of regression residuals.

Meanwhile, Tiefelsdorf and Griffith (2007) outline the general spatial autoregressive model specification foundation for MESF with regard to a random effects term. For this conceptualization, consider the following spatial simultaneous autoregressive (SAR) model (the spatial error model in spatial econometrics) specification:

$$\mathbf{Y}=\mathbf{X}\boldsymbol{\beta}_{\mathbf{X}}+(\mathbf{I}-\rho\mathbf{C})^{-1}\boldsymbol{\varepsilon}. \tag{3.5}$$

The inverse matrix in Eq. (3.2), which relates to geostatistics and spectral density functions (see Chapter 2), may be expanded as follows and is a convergent series when subjected to the restriction $1/\lambda_n < \rho < 1/\lambda_1$, where λ_n and λ_1 are the extreme eigenvalues of the SWM $\mathbf{C}$ (see Bivand 1984, p. 26):

$$(\mathbf{I}-\rho\mathbf{C})^{-1}=\sum_{h=0}^{\infty}\rho^h\mathbf{C}^h.$$

Therefore,

$$\begin{aligned}\mathbf{Y}&=\mathbf{X}\boldsymbol{\beta}_\mathbf{X}+\sum_{k=0}^{\infty}\rho^k\mathbf{C}^k\boldsymbol{\varepsilon}=\mathbf{X}\boldsymbol{\beta}_\mathbf{X}+\rho^0\mathbf{C}^0\boldsymbol{\varepsilon}+\sum_{k=1}^{\infty}\rho^k\mathbf{C}^k\boldsymbol{\varepsilon}\\&=\mathbf{X}\boldsymbol{\beta}_\mathbf{X}+\sum_{k=1}^{\infty}\rho^k\mathbf{C}^k\boldsymbol{\varepsilon}+\boldsymbol{\varepsilon}.\end{aligned}$$

The term $(\sum_{k=1}^{\infty}\rho^k\mathbf{C}^k\boldsymbol{\varepsilon}+\boldsymbol{\varepsilon})$ is the regression residuals, with the subterm $(\sum_{k=1}^{\infty}\rho^k\mathbf{C}^k\boldsymbol{\varepsilon})$ being the source of SA. Its inclusion of exponential powers of the SWM $\mathbf{C}$ explains why the eigenvectors of this matrix, which do not change with the powering operation, can be used to filter SA from linear regression residuals. The equation

$$\mathbf{Y}=\mathbf{X}\boldsymbol{\beta}_\mathbf{X}+\sum_{k=1}^{\infty}\rho^k\mathbf{C}^k\boldsymbol{\varepsilon}+\boldsymbol{\varepsilon} \tag{3.6}$$

is of the random effects model specification form

$$\mathbf{Y}=\mathbf{X}\boldsymbol{\beta}_\mathbf{X}+\mathbf{Z}\boldsymbol{\gamma}+\boldsymbol{\varepsilon},$$

where matrix $\mathbf{Z}$ is an n-by-K design matrix (here it contains eigenvectors), and $\boldsymbol{\gamma}$ is a K-by-1 vector of unknown random effects affiliated with design matrix $\mathbf{Z}$ such that it is a normal RV with a diagonal covariance matrix whose diagonal entries are all the same. For georeferenced data, a term like $\mathbf{Z}\boldsymbol{\gamma}$ comprises a spatially structured (SSRE that is SA based) and a spatially unstructured (SURE) random effects component. Estimating the SSRE component of $\mathbf{Z}\boldsymbol{\gamma}$ usually requires additional information, such as repeated measures furnished by a space–time series or parameter priors in a Bayesian analysis, but the redundant information in linear regression residuals represented by SA allows estimation of this SSRE component with only a single cross-sectional (i.e., georeferenced data for a single point in time) dataset.

In summary, spatial autoregression in general furnishes a foundation for MESF, either in terms of an AR or a SAR model specification. This section outlines the algebra for two such derivations. In doing so, a linkage emerges with random effects model specifications, which first were used extensively in spatial statistics/econometrics in Bayesian map analysis.

3.1.1 The fundamental theorem of MESF

A statement of the fundamental theorem of MESF appears in Section 2.1.3. It is based upon several theorems in matrix algebra, including the fundamental theorem of principal components analysis (see Tatsuoka, 1988, p. 146), which may be translated as follows:

> *Given a modified n-by-n SWM $(\mathbf{I} - \mathbf{11}^T/n)\mathbf{C}(\mathbf{I} - \mathbf{11}^T/n)$ for a given geographic landscape, we can derive a set of orthogonal and uncorrelated variables $\mathbf{E}_1, \mathbf{E}_2, \ldots, \mathbf{E}_n$ by a set of linear transformations corresponding to the principal-axes rotation [i.e., the rigid rotation whose transformation matrix $\mathbf{E}$ has the n eigenvectors of matrix $(\mathbf{I} - \mathbf{11}^T/n)\mathbf{C}(\mathbf{I} - \mathbf{11}^T/n)$] as its columns. The SA measures of this new set of variables are given by the diagonal matrix $(n/\mathbf{1}^T\mathbf{C1})\mathbf{\Lambda} = [n/\mathbf{1}^T\mathbf{C1}]\mathbf{E}^T(\mathbf{I} - \mathbf{11}^T/n)\mathbf{C}(\mathbf{I} - \mathbf{11}^T/n)\mathbf{E}$, whose diagonal elements are the n MCs of the corresponding map patterns produced by the n eigenvectors of matrix $\mathbf{E}$.*

Orthogonality results from the matrix $(\mathbf{I} - \mathbf{11}^T/n)\mathbf{C}(\mathbf{I} - \mathbf{11}^T/n)$ being symmetric (if $\mathbf{C}$ is a symmetric matrix, then $\mathbf{ACA}^T$ is a symmetric matrix). Uncorrelatedness results from the pre- and postmultiplication of matrix $\mathbf{C}$ by the projection matrix $(\mathbf{I} - \mathbf{11}^T/n)$, resulting in a single eigenvector proportional to the n-by-1 vector $\mathbf{1}$, and hence the $n - 1$ other eigenvectors having elements that sum to zero; the numerator of the Pearson product moment correlation coefficient for a pair of different eigenvectors has a cross-product term (e.g., XY) of zero (orthogonality) and a product of two means (each being a sum of the elements of an eigenvector, with at least one of these sums equal to zero) of zero (Griffith 2000b, p. 105). Tiefelsdorf and Boots (1995; Section 2.1.2) prove that the MC for a given eigenvector $\mathbf{E}_j$ is given by $(n/\mathbf{1}^T\mathbf{C1})\lambda_j$. The rank ordering of the Rayleigh quotients

$$(n/\mathbf{1}^T\mathbf{C1})\mathbf{E}^T(\mathbf{I}-\mathbf{11}^T/n)\mathbf{C}(\mathbf{I}-\mathbf{11}^T/n)\mathbf{E}/(\mathbf{E}^T\mathbf{E}) = (n/\mathbf{1}^T\mathbf{C1})\mathbf{\Lambda}$$

produces the sequential ordering from the maximum possible level of positive SA (PSA) to the maximum possible level of negative SA (NSA; see de Jong, Sprenger, & van Veen, 1984).

Because one eigenvector element corresponds to each of the n areal units in a geographic landscape, a map can portray the geographic distribution of each set of eigenvector elements. Consequently, a map of the ESF $\mathbf{E}_K\boldsymbol{\beta}_E$ furnishes a visualization of SA; as such, it supplements the Moran scatterplot graphic tool. Furthermore, because each eigenvector is an n-by-1 variate, eigenvectors can be treated like covariates and included in a linear regression analysis.

3.1.2 Map pattern and SA: Heterogeneity in map-wide trends

SA may be interpreted in a number of different ways, one of which is map pattern (Griffith, 1992). Pattern refers to some discernible real-world regularity that contains elements recurring in a predictable manner. Map pattern refers to this regularity and repetitiveness occurring in two dimensions and is the basis for spatial interpolation (prediction linking to kriging in geostatistics). SA makes map pattern possible by organizing attribute values on a map in such a way that for PSA, for example, relatively high values cluster together in a geographic landscape, as do relatively intermediate, and relatively low, values. This geographic organization can yield global gradients across, as well as large regional or small local clusters in, a geographic landscape; in general, neighborhood subsets of georeferenced attribute values are similar or dissimilar (NSA). These are the components of map pattern depicted by the modified SWM eigenvectors with, respectively, large, moderate, or small but not close to zero, eigenvalues. In other words, map pattern has to do with the geographic arrangement of attribute values of a map, with the nature and degree of (dis)similarities of nearby values relating to SA.

Heterogeneity refers to a collection of diverse elements, elements that are nonuniform in the composition of their attribute values. In terms of statistical properties, these elements are not IID (see Section 3.1). In classical linear regression, a response variable Y often is considered heterogeneous in its individual observation means, resulting in the term $\mathbf{X}\boldsymbol{\beta}_\mathbf{X}$ being included in a linear regression specification. This specification strategy seeks to account for heterogeneity with the regression mean, rendering residuals that are IID and hence homogeneous. If $\mathbf{X} \equiv \mathbf{1}$, then the mean of Y for each areal unit is the constant β_0; this is the special case of a homogeneous Y. In the presence of SA, the residuals still have a mean of zero, but now heterogeneity persists through their variances being unequal; this outcome is one consequence of variance inflation by SA. Eq. (3.4) highlights how MESF addresses this problem by replacing the constant mean with a variable mean:

$$\mathbf{Y} = \mathbf{E}_K\boldsymbol{\beta}_\mathbf{E} + \mathbf{X}\boldsymbol{\beta}_\mathbf{X} + \xi = (\mathbf{1}\beta_0 + \mathbf{E}_K\boldsymbol{\beta}_\mathbf{E}) + \mathbf{X}_p\boldsymbol{\beta}_\mathbf{X} + \xi,$$

where the vector $\mathbf{1}$ has been removed from the n-by-(p+1) $\mathbf{X}$ matrix, denoted by $\mathbf{X}_p$. The varying intercept term is $(\mathbf{1}\beta_0 + \mathbf{E}_K\boldsymbol{\beta}_\mathbf{E})$. In random effects specifications, a distributional assumption is attached to this composite variance. In keeping with classical linear regression theory, an argument can be made here that an appropriate distributional assumption is normality (see Bentkus, Bloznelis, & Götze, 1996; Chaidee and Tuntapthai (2009):

$K << n$ eigenvectors extracted from matrix $(\mathbf{I} - \mathbf{11}^T/n)\mathbf{C}(\mathbf{I} - \mathbf{11}^T/n)$ are mutually independent and are not necessarily identically distributed RVs. As K goes to infinity, the linear combination ESF constructed with them converges on a normal distribution.

The critical feature here is that a linear regression intercept is allowed to vary, resulting in the response variable value Y for each individual areal unit being predictable by a heterogeneous intercept term that varies across a geographic landscape and most likely conforms to a bell-shaped curve. Therefore a constructed ESF describes heterogeneity in map-wide trends. This heterogeneity may take the form of a landscape-wide attribute value gradient for global map-wide trends, regional contrasts for relatively large clusters of similar attribute values, and local contrasts for relatively small clusters of similar attribute values.

Griffith (2008) discusses how MESF also can describe attribute variable regression slopes varying across a geographic landscape because of SA. A slope ESF, just like an intercept ESF, displays individual areal unit deviations from a global value.

3.2 Estimating an ESF as an OLS problem: An illustrative linear regression example

A pure SA analysis ignores covariates and estimates the SA latent in an attribute variable map pattern. This type of analysis is pertinent to, for example, the construction of histograms for georeferenced RVs or the calculation of Pearson product moment correlation coefficients for pairs of georeferenced RVs, among other things. The MESF linear regression equation, which assumes normally distributed residuals, is the following nonconstant mean-only specification:

$$\mathbf{Y} = \mathbf{1}\beta_0 + \mathbf{E}_K\boldsymbol{\beta}_\mathbf{E} + \boldsymbol{\xi}. \tag{3.7}$$

One traditional specification error concern here pertains to how closely the response variable Y conforms to a normal distribution. Analysts frequently subject a nonnormal set of attribute values to a Box–Cox/Manly transformation to normality (see Griffith, 2013).

Consider the 2010 population density (PD) across the 254 counties of Texas (see Fig. 3.1A); urban areas are conspicuous in this map pattern, revealing geographic heterogeneity. Raw PD values do not conform closely to a bell-shaped curve (Fig. 3.2A), whereas Box–Cox transformed values LN(PD − 0.08) do (Fig. 3.2B), where LN denotes natural logarithm.

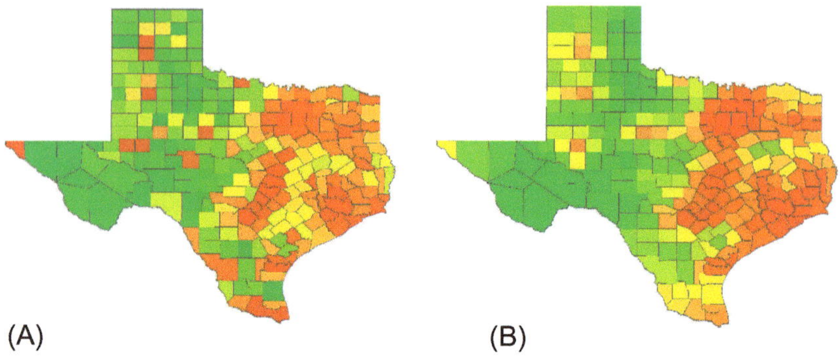

Fig. 3.1 2010 PD by county across Texas; PD and ESF values increase from *dark green* to *yellow* to *dark red*. (A) The geographic distribution of Box–Cox transformed values. (B) The linear regression ESF component of Box–Cox transformed PD.

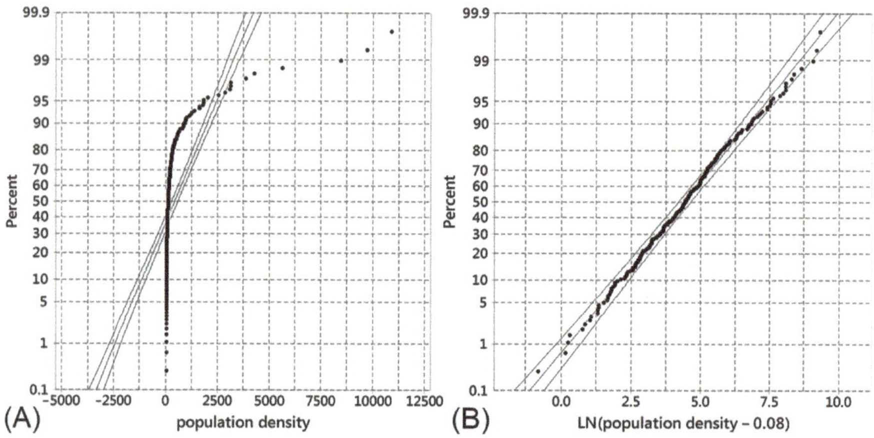

Fig. 3.2 Normal quantile plots with 95% confidence intervals for 2010 PD by county across Texas; *lines* denote theoretical results, and *black dots* denote PD values. (A) The raw PD values. (B) The Box–Cox log-transformed PD values.

The Shapiro–Wilk normality diagnostic statistic for the raw PD variable is 0.33644 ($p < 0.0001$), whereas it is 0.99600 ($p = 0.7624$) for the transformed PD variable. For initial analysis purposes, then, Y = LN(PD − 0.08), and the linear regression equation of interest is given by

$$\mathbf{Y} = \mathbf{1}\beta_0 + (\mathbf{E}_K\boldsymbol{\beta}_\mathbf{E} + \boldsymbol{\xi}),$$

which involves calculating $\mathbf{Y} = \mathbf{1}b_0 + \mathbf{e}$ and then evaluating the n-by-1 vector of regression residuals $\mathbf{e}$ for SA. For this example, $b_0 = 4.40986$, and the SA index analyses are as follows:

$$MC = 0.52176, z_{MC} = 13.52 \text{ based upon the asymptotic}$$

$$\sigma_{MC} \approx \sqrt{(2/\mathbf{1}^T\mathbf{C}\mathbf{1})} = \sqrt{2/1322}.$$

Geary ratio (GR) = 0.44315, $z_{GR} = -5.60$ based upon the asymptotic

$$\sigma_{GR} \approx \sqrt{\frac{2}{\mathbf{1}^T\mathbf{C}\mathbf{1}} + 2\frac{\mathbf{1}^T\mathbf{C}^2\mathbf{1}}{(\mathbf{1}^T\mathbf{C}\mathbf{1})^2}} = \sqrt{\frac{2}{1322} + 2\frac{7304}{(1322)^2}}.$$

The implication from both SA indices is that the transformed PD map pattern contains moderate PSA. An ESF needs to be constructed to account for this residual SA.

3.2.1 The selection of eigenvectors to construct an ESF

The first step in constructing an ESF for the 2010 Texas PD by county is to extract the 254 eigenvectors from the modified SWM $(\mathbf{I} - \mathbf{1}\mathbf{1}^T/254) \times \mathbf{C}(\mathbf{I} - \mathbf{1}\mathbf{1}^T/254)$, where 0–1 matrix $\mathbf{C}$ denotes the Texas county SWM, based upon the rook definition of adjacency (see Preface, Fig. P1). Because this PD exhibits PSA, determining an appropriate candidate set of eigenvectors for stepwise regression can begin by setting aside the 149 NSA eigenvectors plus the single eigenvector having a zero eigenvalue (corresponding to the eigenvector proportional to the vector $\mathbf{1}$), which a regression equation already includes for its intercept term. The next step is to determine how many of the 104 PSA eigenvectors to include, counting this number from the largest eigenvalue (i.e., the maximum possible PSA). Chun et al. (2016, p. 75) furnish the following equation to help with this decision:

$$\frac{n_{pos}}{1 + \exp\left\{2.1480 - \frac{6.1808(z_{MC} + 0.6)^{0.1742}}{n_{pos}^{0.1298}} + \frac{3.3534}{(z_{MC} + 0.6)^{0.1742}}\right\}},$$

with $n_{pos} = 104$ (the number of PSA eigenvectors), and $z_{MC} = 13.52$ (the linear regression residuals z-score measure of SA) here. This expression indicates that the candidate set should contain the 78 eigenvectors with the largest eigenvalues.

One useful criterion for eigenvector selection from the candidate set is the level of significance for each eigenvector's regression coefficient, which essentially maximizes the linear regression R^2 value; other selection criteria could be utilized (see Griffith, 2004). In addition, a stepwise procedure that combines both forward selection and backward elimination supports the construction of a parsimonious ESF. Because the eigenvectors are mutually orthogonal and uncorrelated, the primary factor in eigenvector selection during any given step is the marginal error sum of squares for that step. Of the 78 candidate eigenvectors, 26 were selected using a significance level criterion of 0.10, accounting for roughly 62.5% of the variation in log-transformed PD across the counties of Texas (Fig. 3.1B), highlighting the Dallas, Houston, and Austin–San Antonio metropolitan regions and indicating that SA introduces variance inflation by more than doubling the underlying IID variance. Table 3.1 summarizes the stepwise selection results, revealing that global (e.g., $\mathbf{E}_2$), regional (e.g., $\mathbf{E}_{18}$), and local (e.g., $\mathbf{E}_{77}$) map pattern[1] components account for the SA under study and that the degree of SA does not determine the selection sequence.

3.2.2 Selected criteria for assessing regression models: The PRESS statistic, residual diagnostics, and multicollinearity

Once an ESF is constructed, model diagnostics should be performed. The predicted residual error sum of squares (PRESS) statistic is a useful global diagnostic to calculate because it relates to a cross-validation assessment, with the set of covariates being held constant. Values of the ratio PRESS/ESS close to 1, where ESS denotes error sum of squares, indicate good model performance in this context because the corresponding estimated model fitting and prediction error essentially are the same (i.e., the estimated trend line also describes new observations well). Here this values is 376.737/355.645 = 1.059, implying a very respectable model performance with regard to the cross-validation criterion.

Three features of the linear regression residuals merit assessment. The first concerns normality (Fig. 3.3A); here the Shapiro–Wilk statistic for the linear regression residuals is 0.98030 ($p = 0.0014$); the frequency distribution for these residuals differs statistically, but not substantively, from a bell-shaped curve. The second concerns residual SA. The expected value

[1]The grouping into global, regional, and local map patterns is subjective. These terms, respectively, refer to MC/MC_{max} (i.e., the maximum MC) values in the ranges 0.9–1, 0.7–0.9, and 0.25–0.7. The maximum MC value here is 1.09798, which should be used to standardize MC values to make them comparable across geographic landscapes.

Table 3.1 Eigenvectors selected to account for SA in the geographic distribution of Box–Cox transformed 2010 PD by county across Texas

Step	Eigenvector entered	MC	Regression coefficient	Partial R^2	Model R^2	Selection probability	Simulation probability #1	Simulation probability #2	Simulation probability #3
1	E_{3}	1.05566	10.9929	0.1487	0.1487	<0.0001	**1.0000**	**1.0000**	**1.0000**
2	E_{2}	1.07484	−7.0831	0.0617	0.2104	<0.0001	**1.0000**	**1.0000**	**1.0000**
3	E_{13}	0.86872	−6.5413	0.0527	0.2631	<0.0001	**1.0000**	**1.0000**	**1.0000**
4	E_{1}	1.09798	6.5373	0.0526	0.3157	<0.0001	**1.0000**	**1.0000**	**1.0000**
5	E_{25}	0.69942	6.2470	0.0480	0.3637	<0.0001	**1.0000**	**1.0000**	**1.0000**
6	E_{8}	0.94955	5.2157	0.0335	0.3972	0.0003	**0.9976**	**0.9994**	**0.9996**
7	E_{4}	1.03282	−4.7126	0.0273	0.4245	0.0007	**0.9942**	**0.9965**	**0.9960**
8	E_{56}	0.33756	−4.5687	0.0257	0.4502	0.0008	**0.9906**	**0.9965**	**0.9934**
9	E_{9}	0.93736	4.3500	0.0233	0.4735	0.0012	**0.9847**	**0.9918**	**0.9833**
10	E_{23}	0.72244	3.7131	0.0170	0.4905	0.0048	**0.9417**	**0.9656**	**0.9866**
11	E_{18}	0.79174	−3.3633	0.0139	0.5044	0.0097	0.8999	**0.9292**	**0.9147**
12	E_{35}	0.56992	3.3229	0.0136	0.5180	0.0097	0.8910	**0.9238**	**0.9553**
13	E_{52}	0.39483	−3.3084	0.0135	0.5314	0.0092	0.8938	**0.9231**	**0.9308**
14	E_{28}	0.66591	−3.2867	0.0133	0.5447	0.0088	0.8903	**0.9242**	**0.9421**
15	E_{77}	0.17989	2.9286	0.0106	0.5553	0.0183	0.8262	0.8603	**0.9480**
16	E_{11}	0.89249	−2.7316	0.0092	0.5645	0.0263	0.7745	0.8197	0.8216
17	E_{75}	0.19355	−2.7007	0.0090	0.5734	0.0268	0.7740	0.8111	0.8474
18	E_{43}	0.47611	2.3185	0.0066	0.5801	0.0556	0.6592	0.6998	0.6969
19	E_{21}	0.75694	2.2571	0.0063	0.5863	0.0609	0.6350	0.6775	0.7510
20	E_{15}	0.82277	−2.2358	0.0062	0.5925	0.0620	0.6341	0.6707	0.6279
21	E_{61}	0.30956	−2.1948	0.0059	0.5984	0.0655	0.6189	0.6652	0.7261
22	E_{10}	0.90190	2.1559	0.0057	0.6041	0.0690	0.6051	0.6552	0.8103
23	E_{38}	0.53291	−2.1338	0.0056	0.6097	0.0705	0.6069	0.6361	0.6070
24	E_{14}	0.84671	2.0592	0.0052	0.6149	0.0795	0.5795	0.6110	0.6048
25	E_{27}	0.67213	2.0237	0.0050	0.6200	0.0834	0.5709	0.5985	0.6727
26	E_{42}	0.48766	1.9489	0.0047	0.6247	0.0941	0.5336	0.5731	0.5953
Others							0.1082	0.1094	0.1092

Note: The simulation probabilities are based upon 10,000 replications; simulation probabilities >0.9 are in bold.

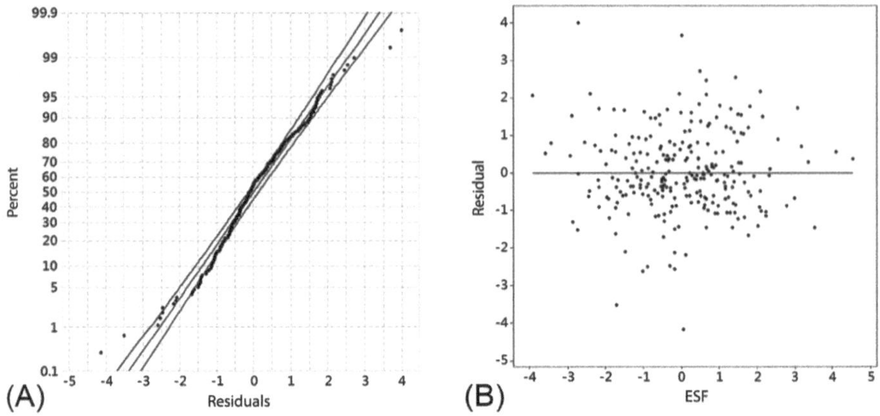

Fig. 3.3 Texas 2010 log-transformed PD ESF linear regression residuals graphical diagnostics. (A) A normal quantile plot. (B) An equality of residuals variance plot.

of the MC for linear regression residuals calculated for an ESF-only regression is:

$$E(MC) = -\frac{1 + \sum_{j=1}^{K} MC_j}{n - K - 1},$$

where MC_j is the MC for the jth eigenvector included in a regression equation. If no eigenvectors are selected, then this expression reduces to $-1/(n-1)$, the standard result for a single attribute variable. Here the value is (the MC_js appear in Table 3.1)

$$(1 + 18.27137)/(254 - 26 - 1) = -0.08490.$$

Accordingly, the residual z_{MC} is $(-0.05177 + 0.08490)/\sqrt{2/1322} = 0.85$ (a dramatic reduction from 13.52), implying that the residuals contain only a trace amount of PSA. The GR can be rewritten as a function of the MC:

$$GR = \frac{n-1}{\sum_{i=1}^{n}\sum_{j=1}^{n} c_{ij}} \frac{\sum_{i=1}^{n} (y_i - x_i b)^2 \left(\sum_{j=1}^{n} c_{ij}\right)}{\sum_{i=1}^{n} (y_i - x_i b)^2} - \frac{n-1}{n} MC.$$

Therefore $E(GR) = \frac{n-1}{n}\left[1 - \frac{1 + \sum_{j=1}^{K} MC_j}{n-K-1}\right]$. The GR result corroborates the MC result (its z-score is negative because MC and GR are inversely related):

$$z_{GR} = (0.97567 - 1.08063)/0.09935 = -1.06.$$

The third residual feature concerns heteroscedasticity (i.e., attribute variance heterogeneity); Fig. 3.3B fails to disclose any conspicuous pattern in the scatterplot of the residuals versus their corresponding means, implying homogeneity of variance.

The third diagnostic concerns multicollinearity. The ESF eigenvectors are mutually orthogonal and uncorrelated. Thus their variance inflation factors (VIFs) all should be 1 in the pure SA case. Indeed, here they all are 1.

Finally, as an aside, with regard to viewing an ESF as a SSRE, each component should have the same variance, which it does here (1.15917); the ratio of the maximum and minimum calculated ESF value variances is 1.16, which is not significantly >1. Also, the ESF conforms to a bell-shaped curve (Shapiro–Wilk = 0.99706, p = 0.9236). In other words, the ESF term satisfies many of the requirements of a SSRE term.

3.2.3 Interpreting an ESF and its parameter estimates

As mentioned previously, the constructed ESF (Fig. 3.1B) highlights the major urban regions of Texas. It is a smoothed version of its empirical counterpart (Fig. 3.1A). One interpretation criterion for it pertains to its global SA index values: MC = 0.86603, GR = 0.19205. In other words, this ESF portrays moderate to strong PSA, which is characteristic of regional map patterns such as those representing multiple dispersed metropolitan areas in a geographic landscape. As mentioned in the preceding section, this ESF satisfies selected characteristics of a SSRE term. In addition, its PRESS statistic indicates that it furnishes respectable cross-validation results, suggesting that it could serve as a reasonable interpolator (see the Moran scatterplot in Fig. 3.4A).

The principal interpretation of the constructed ESF is that high population densities tend to concentrate in urban regions, and low population densities tend to concentrate in rural regions. Because Texas contains a number of major urban regions, and central place theory implies that they should be dispersed across the state, the PD map pattern reflects regional clusters rather than a global PD gradient across this geographic landscape. This geospatial characteristic results in a MC value in the top quintile, rather than nearly at the top, of possible degrees of PSA. This outcome also implies that cities should be conspicuous in the ESF map pattern. This visualization detail supplements the explanation furnished by a global SA index, even when a Moran scatterplot augments it.

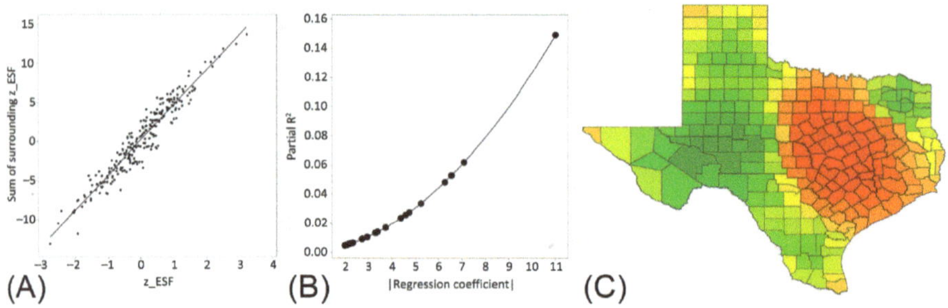

Fig. 3.4 Graphical portrayals of ESF features. (A) The Moran scatterplot. (B) The relationship between the regression coefficients and the percentages of variance accounted for by the individual eigenvectors. (C) The geographic distribution of the most prominent eigenvector ($\mathbf{E}_3$); vector elements increase in value from *dark green* to *yellow* to *dark red.*

The regression coefficients used to construct the linear combination of eigenvectors that constitutes the 2010 Texas log-transformed PD ESF appear in Table 3.1. Because an eigenvector is unique except for a multiplicative factor of −1, the signs of the coefficients are irrelevant. Rather, their absolute magnitudes indicate their relative importance. The relationship between the partial R^2s and the absolute magnitude of the ESF regression coefficients is quadratic (see Fig. 3.4B). The eigenvectors for the largest regression coefficients tend to dominate the ESF. The most prominent eigenvector portrays an east–west contrast (Fig. 3.4C), highlighting the part of the state containing Dallas, Houston, and Austin–San Antonio. The remaining 25 eigenvectors tend to capture details that convert Fig. 3.4C to Fig. 3.1B (notice the sizable gap between the first—that is, the maximum horizontal/vertical axes value—and the other regression coefficients in Fig. 3.4B).

3.2.4 Comparisons between ESF and SAR model specification results

The simplest version of MESF accounts for SA by including a nonconstant mean in a regression model. The spatial SAR specification does this as well by including the term $[(1-\rho)\beta_0\mathbf{1}+\rho\mathbf{WY}]$, where β_0 denotes the intercept term. The pure SA SAR model is specified as

$$\mathbf{Y}=(1-\rho)\beta_0\mathbf{1}+\rho\mathbf{WY}+\boldsymbol{\varepsilon},$$

employing the row-standardized version of matrix $\mathbf{C}$, namely, matrix $\mathbf{W}$. For the Box–Cox transformed PD studied in this chapter, the maximum likelihood estimate of the SA parameter is $\hat{\rho}=0.70120$. This SA term accounts for about 48.1% of the variance in the Box–Cox transformed PD across Texas. This percentage is less than the 62% for the ESF specification, in part because the SAR specification includes all, not only the relevant subset of, eigenvectors, introducing some noise into its estimation. Meanwhile, the SAR residual Shapiro–Wilk statistic, 0.96204, is statistically significant ($p < 0.0001$). Both Getis and Griffith (2002) and Thayn and Simanis (2013) present comparisons of spatial autoregressive and ESF analyses. An ESF specification frequently outperforms a spatial autoregressive specification.

Perhaps one of the greatest advantages MESF has vis-à-vis spatial autoregression is its ability to visualize the SA latent in a georeferenced attribute variable. It also has implementation advantages for generalized linear models (GLMs; see Chapter 5).

3.3 Simulation experiments based upon ESFs

Griffith (2017) argues that MESF is superior to spatial autoregression for spatial statistical simulation experiments because it preserves an underlying map pattern and is characterized by constant variance; in other words, it supports conditional geospatial simulations. A spatial analyst can undertake a simulation experiment employing MESF in one of the following three ways: (1) draw a random error term from a normal distribution with mean zero and variance equal to the linear regression mean squared error; (2) randomly permute the n residuals calculated with linear regression estimation; and, (3) randomly sample, with replacement, the n residuals from the linear regression estimation (similar to bootstrapping). Each of these three strategies was used to perform a sensitivity analysis simulation for the ESF constructed in Section 3.2.2. Each simulation experiment involved 10,000 replications (to profit from the Law of Large Numbers).

The first simulation experiment added random noise $\varepsilon_i \sim N(0, 1.24350^2)$, $i = 1, 2, \ldots, 254$, to the ESF + intercept term (i.e., 4.40986). The simulation mean of the map averages (based upon sets of 254 ε_i) is −0.00045; the simulation mean of the map variances is 1.15826. Fig. 3.5A portrays the simulated mean map pattern for the simulated log-transformed PD values; it essentially is identical to the map pattern in Fig. 3.1B. The variances for the individual county simulations span the range from 1.13704^2 to 1.18137^2; the F-ratio for these two extreme variances is 1.08, which is not statistically significant, yielding a single variance class (Fig. 3.5B). One important advantage of MESF vis-à-vis spatial autoregression-based simulation experiments is that the variance is constant across a geographic landscape, which is not the case

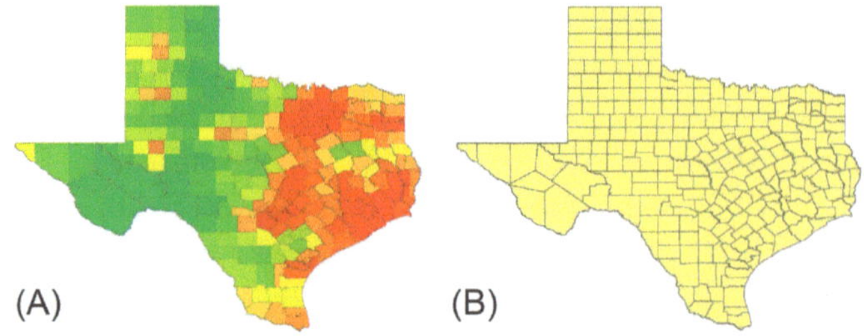

Fig. 3.5 Output from simulation experiment #1. (A) The average map pattern; county simulation averages increase from *dark green* to *yellow* to *dark red*. (B) County-specific variances of the average map pattern based upon the single class interval [1.13, 1.17].

for spatial autoregression (see Griffith, 2017). The simulation mean R^2 value is 0.6699, which is somewhat greater than the actual R^2 value. Meanwhile, the simulation mean Shapiro–Wilk probability is 0.50136.

Table 3.1 tabulates the eigenvector selection significance level probabilities, p_{sig}, as well as the eigenvector selection simulation probabilities, p_{sim}. Using a 10% level of significance selection criterion renders roughly a 10% chance that some of the 52 eigenvectors not selected in the original analysis are selected in a simulation analysis. The relationship between these two selection probabilities may be described as follows:

$$0.24\left(e^{-3.35p_{sim}^{2.2}} - e^{-3.35}\right), \text{pseudo-}R^2 \approx 1.0000.$$

This description confirms a close relationship between the original selection and the simulation selection probabilities.

The second simulation experiment resamples from the original linear regression residuals, sampling without replacement. Each sample is a permutation of the original residuals across the 254 counties. These new residuals are added to the ESF + intercept term (i.e., 4.40986). For every simulation replicate, the map mean and variance are identical, respectively, equaling zero and the mean squared error for the original linear regression analysis. The summary statistics are very similar to those produced by the first simulation experiment: an extreme variances F-ratio of 1.11, a simulation mean R^2 of 0.69209, and a simulation mean Shapiro–Wilk probability of 0.1812. In addition, the simulation selection probabilities also are very similar to those produced by the first experiment (see Table 3.1). Finally, the simulation mean and variance maps are almost identical to those appearing in Fig. 3.5.

The third simulation experiment resamples from the original linear regression residuals, sampling with replacement; this is a bootstrap resampling design. This strategy produces results that differ noticeably from those produced by the other two simulation designs. The simulation selection probabilities disagree to some degree for a number of the eigenvectors. The R^2 still is high (i.e., 0.69447). However, the average Shapiro–Wilk probability is much smaller (i.e., 0.1605), and the extreme variances F-ratio of 6.85 implies nonconstant variance across the state of Texas. The simulation mean map also differs from that portrayed in Fig. 3.5A (see Fig. 3.6A). Resampling with replacement from a set of original linear regression residuals appears to be an inferior simulation experimental design.

In summary, MESF furnishes an appealing basis for spatial statistical simulation experiments because it preserves an underlying map pattern. This part of a simulation design should be coupled with sampling errors either from the assumed theoretical normal distribution [what Cliff & Ord (1973), call free sampling] or from the set of calculated linear regression residuals by permuting them (i.e., sampling without replacement, with order being important).

3.4 ESF prediction with linear regression

Prediction is a valuable use of linear regression and is alluded to by the PRESS statistic. Redundant attribute information (i.e., multicollinearity) with the covariates supports the prediction of the response variable; each of these predictions is a conditional mean (i.e., a regression fitted value) based upon the given covariates used to compute it. An extension of this prediction capability is to observations not included in the original sample; a set of estimated regression coefficients enables the calculation of a prediction with covariates measured for out-of-sample observations. These supplemental observations have an additional source of variation affiliated with them, namely, their own stochastic noise, which is not addressed during estimation of the already-calculated regression coefficients.

Cross-validation offers an application of ESF prediction with linear regression. This prediction may be executed with the following modified pure SA linear regression specification when a single attribute variable value, y_m, is missing:

$$\begin{pmatrix} \mathbf{Y}_o \\ 0 \end{pmatrix} = \beta_0 \mathbf{1} - y_m \begin{pmatrix} \mathbf{0}_o \\ 1 \end{pmatrix} + \sum_{k=1}^{K} \begin{pmatrix} \mathbf{E}_{o,k} \\ \mathbf{E}_{m,k} \end{pmatrix} \boldsymbol{\beta}_{E_K} + \begin{pmatrix} \boldsymbol{\varepsilon}_o \\ 0 \end{pmatrix}, \qquad (3.8)$$

where the subscript o denotes observed data, the subscript m denotes missing data, and $\mathbf{0}$ is a vector of zeros. This specification subtracts the unknown data

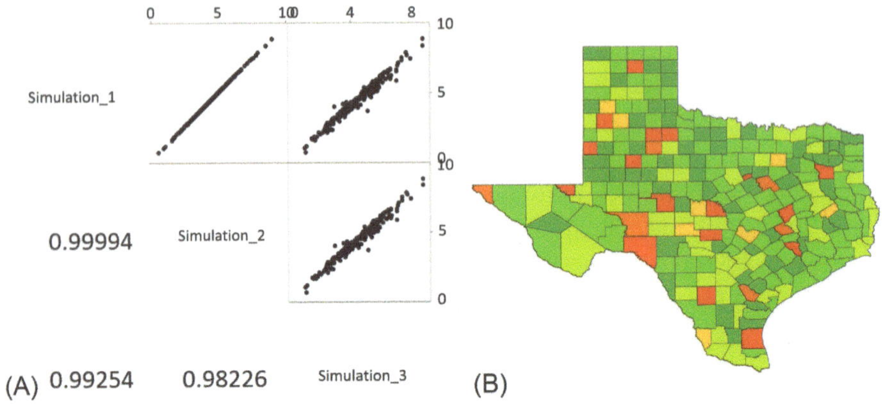

Fig. 3.6 Output from simulation experiment #3. (A) Scatterplots (*upper triangle*) and correlation coefficients (*lower triangle*) for the average map patterns from the three simulation experiments. (B) Variance of the average map pattern based upon six equal-sized intervals [0.68, 1.80]; variance increases from *dark green* to *yellow* to *dark red*.

values, y_m, from both sides of the equation and then allows these values to be estimated as regression parameters (i.e., conditional means). In doing so, these conditional means are equivalent to their fitted values and hence have residuals of zero.

Fig. 3.7 portrays the scatterplot of the log-transformed 2010 Texas PD (vertical axis) versus the corresponding 254 imputed values calculated with Eq. (3.8) but with no covariates (i.e., a pure SA specification); this exercise is similar to kriging. The linear regression equation describing this correspondence may be written as follows:

$$\hat{Y} = 0.98849 + 0.77990Y_{\text{predicted}}, R^2 = 0.4078.$$

Its residual Shapiro–Wilk probability is <0.0001.

Fig. 3.7B furnishes a tabulation of the number of individual estimated ESF linear regression equations that either failed to select some of the original 26 eigenvectors or selected some of the remaining 52 eigenvectors not included in the original estimated linear regression equation. Eigenvector $\mathbf{E}_{42}$ is not selected a substantial number of times (60/254). In contrast, perhaps eigenvector $\mathbf{E}_{36}$ is selected a relatively large number of times (25/254) for not being included in the original ESF linear regression equation. Overall, the correct set of eigenvectors is selected for each of the n subsets of size $n - 1$ used to compute imputations.

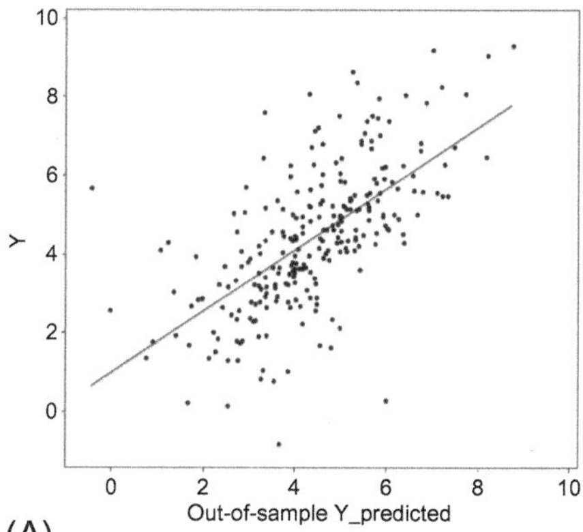

(A)

Eigen-vector	Missed selections	Eigen-vector	Extraneous selections
E_{10}	5	E_{17}	1
E_{14}	10	E_{20}	4
E_{15}	3	E_{24}	2
E_{21}	1	E_{26}	1
E_{27}	17	E_{29}	5
E_{38}	4	E_{31}	4
E_{42}	60	E_{33}	5
E_{43}	1	E_{36}	25
E_{61}	4	E_{39}	2
		E_{45}	2
		E_{60}	1

(B)

Fig. 3.7 Selected out-of-sample predicted results. (A) A scatterplot of the log-transformed PD values (Y) versus their out-of-sample predicted values. (B) A tabulation of selection failure and extraneous eigenvectors.

3.5 Summary

This chapter presents a theoretical foundation for MESF in the context of a normal (i.e., Gaussian) RV and classical linear regression. It summarizes justification for the fundamental theorem of MESF stated in Chapter 2. It highlights problems with and solutions for the presence of heterogeneity in georeferenced data, relating this heterogeneity to map pattern and SA. Also, it demonstrates how to construct an ESF using linear regression and OLS estimation, and then how to evaluate its quality with selected diagnostics. The primary approach presented for ESF construction is a procedure for the stepwise selection of regression covariates; implementing this strategy requires treating eigenvectors as covariates. Once constructed, an ESF needs to be interpreted; this interpretation is based upon its MC and GR indices, its range of prediction error variance, the statistical distribution it mimics, and its ability to replicate the map pattern displayed by its corresponding response variable Y. A nonlinear relationship also has been uncovered between the marginal percentages of variance accounted for by individual eigenvectors and their matching regression coefficients. Because so much of mathematical spatial statistics is analytically intractable, simulation experiments play an important role in its theoretical development. One advantage of MESF is that it preserves a given map pattern throughout a simulation experiment, and it is capable of maintaining constant variance across areal units, too. This constant variance property is not the case for spatial autoregression-based simulation experiments. Furthermore, either sampling residuals from a particular theoretical statistical distribution or, alternatively, randomizing calculated residuals over a given set of areal units, appear to be better than bootstrapping for simulation experiment replication. This may not be the case for block bootstrapping (Lahiri, 2003). Finally, MESF supports predication, certainly in the tradition of kriging, connecting it to the expectation–maximization (E-M) algorithm developed in the statistical sciences. Overall, this chapter establishes the fundamental foundation for implementing MESF, building on the existing large body of normal curve theory, and sets the stage for subsequent chapters in this book.

References

Bentkus, V., Bloznelis, M., & Götze, F. (1996). A Berry-Esséen bound for Student's statistic in the non-I.I.D. Case. *Journal of Theoretical Probability, 9*, 765–796.

Bivand, R. (1984). Regression modeling with spatial dependence: an application of some class selection and estimation methods. *Geographical Analysis, 16*, 25–37.

Chaidee, N., & Tuntapthai, M. (2009). Berry-Esséen bounds for random sums of non-I.I.D. random variables. *International Mathematical Forum, 4*, 1281–1288.

Chun, Y., Griffith, D., Lee, M., & Sinha, P. (2016). Eigenvector selection with stepwise regression techniques to construct eigenvector spatial filters. *Journal of Geographical Systems, 18*, 67–85.

Cliff, A., & Ord, J. (1973). *Spatial autocorrelation*. London: Pion.

Cochrane, D., & Orcutt, G. (1949). Application of least squares regression to relationships containing auto-correlated error terms. *Journal of the American Statistical Association, 44*, 32–61.

de Jong, P., Sprenger, C., & van Veen, F. (1984). On extreme values of Moran's I and Geary's C. *Geographical Analysis, 16*, 17–24.

Getis, A., & Griffith, D. (2002). Comparative spatial filtering in regression analysis. *Geographical Analysis, 34*, 130–140.

Gorroochurn, P. (2015). On Galton's change from "reversion" to "regression". *American Statistician, 70*, 227–231.

Griffith, D. 1978. The impact of configuration and spatial autocorrelation on the investigation and interpretation of geographical models. PhD dissertation, Department of Geography, University of Toronto.

Griffith, D. (1992). What is spatial autocorrelation? Reflections on the past 25 years of spatial statistics. *L'Espace Géographique, 21*, 265–280.

Griffith, D. (1996). Spatial autocorrelation and eigenfunctions of the geographic weights matrix accompanying geo-referenced data. *Canadian Geographer, 40*, 351–367.

Griffith, D. (2000a). A linear regression solution to the spatial autocorrelation problem. *Journal of Geographical Systems, 2*, 141–156.

Griffith, D. (2000b). Eigenfunction properties and approximations of selected incidence matrices employed in spatial analyses. *Linear Algebra and Its Applications, 321*, 95–112.

Griffith, D. (2004). Distributional properties of georeferenced random variables based on the eigenfunction spatial filter. *Journal of Geographical Systems, 6*, 263–288.

Griffith, D. (2008). Spatial filtering-based contributions to a critique of geographically weighted regression (GWR). *Environment and Planning A*, *40*, 2751–2769.
Griffith, D. (2013). Better articulating normal curve theory for introductory mathematical statistics students: power transformations and their back-transformations. *American Statistician*, *67*, 157–169.
Griffith, D. (2017). Some robustness assessments of Moran eigenvector spatial filtering. *Spatial Statistics*, *22*, 155–179.
Lahiri, S. (2003). *Resampling methods for dependent data.* New York: Springer.
Tatsuoka, M. (1988). [with contributions by P. Lohnes] *Multivariate analysis.* New York: Macmillan.
Thayn, J., & Simanis, J. (2013). Accounting for spatial autocorrelation in linear regression models using spatial filtering with eigenvectors. *Annals of the AAG*, *103*, 47–66.
Tiefelsdorf, M., & Boots, B. (1995). The exact distribution of Moran's I. *Environment and Planning A*, *27*(6), 985–999.
Tiefelsdorf, M., & Griffith, D. (2007). Semi-parametric filtering of spatial autocorrelation: the eigenvector approach. *Environment and Planning A*, *39*, 1193–1221.

CHAPTER 4

Software implementation for constructing an ESF, with special reference to linear regression

Chapter outline

Although Griffith and Chun (2014) furnish computer code scripts for implementing Moran eigenvector spatial filtering (MESF) with a range of computer programs, a standard widely distributed software package did not exist untill now that implements MESF. Rather, students and researchers interested in the MESF methodology relied on several public domain implementations on such platforms as MATLAB and R, with R being the most popular. The R package *spdep* provides two functions for eigenvector selection. The *SpatialFiltering* function applies to ordinary linear regression and selects the subset of eigenvectors of the modified spatial weights matrix (SWM) $(\mathbf{I}-\mathbf{11}^{\mathrm{T}}/\mathrm{n})\mathbf{C}(\mathbf{I}-\mathbf{11}^{\mathrm{T}}/\mathrm{n})$ in such a way that the residual Moran coefficient's (MC's) z-score falls within a preset tolerance interval of zero. In contrast, the *ME* function performs this selection for generalized linear models (GLMs) using a permutation-based test. Both functions provide eigenvector selection services but leave considerable work to be done by a user for typical applications of MESF, such as regression modeling and spatial autocorrelation (SA) analysis. The newly published package *spmoran* (Murakami, 2018) performs the entire process of MESF linear regression

Spatial Regression Analysis Using Eigenvector Spatial Filtering
https://doi.org/10.1016/B978-0-12-815043-6.00004-5

modeling, with the regular and random effects specifications as well as unconditional quantile regressions. In addition, *spmoran* facilitates both contiguity- and distance-based SWMs. These R functions make completion of the MESF-based analysis workflow possible, provided a user is familiar with scripting in the R environment.

This chapter presents a user-friendly implementation of MESF, named the *ESF Tool*, in Microsoft Windows. Its fundamental structure is an integration of DotSpatial and R using C#. It is an abridged version of Spatial Analysis using ArcGIS Engine and R (SAAR) by Koo, Chun, and Griffith (2018) with a replacement of ArcGIS Engine components by DotSpatial; in other words, it does not require proprietary software components. It supports MESF functions for linear, Poisson, and binomial regression as well as standard geographic information systems (GIS) functions in the Windows environment. This chapter begins by introducing the framework of the *ESF Tool* and then illustrates its functionality through a typical workflow, which should facilitate replication by a user of the examples in this book and application of the MESF methods present here to a user's own datasets.[1] This chapter then discusses the complexity challenge in eigenvector selection as geographic scale and spatial resolution increase. A divide-and-conquer strategy for eigenvector-based analysis is evaluated for remotely sensed data and a solution is provided to remedy the accuracy issues in eigenfunction calculations by commonly used software libraries.

4.1 Software implementation

MESF functions are implemented in popular statistical environments, including MATLAB, R, and SAS. Although these functions provide a high level of flexibility and performance, end users face a steep learning curve unless they already are familiar with one or more of these statistical environments. Hence, a Windows application, named the *ESF Tool*, was developed to support end users so that they can conduct MESF in a Windows environment with an easy-to-use graphical user interface (GUI).[2]

The *ESF Tool* was developed with C# in a .NET framework and utilizes the two main components R and DotSpatial. It integrates statistical functions in R and GIS functions in DotSpatial using R.NET, an interoperability

[1] Chapter 10 furnishes a toy dataset and output from additional demonstrations utilizing SAAR that supplement the contents of this chapter.

[2] Currently it is available at https://github.com/esftool/esftool, and SAAR is available at http://TheSAAR.github.io.

bridge program to connect R from a .NET environment (R.NET, 2015); that is, internally the MESF functions in the *ESF Tool* pass R code to R and then receive results from R. These results can be stored in a GIS data file[3] or portrayed with a map or both. Its GIS functions rely solely on DotSpatial. The *ESF Tool* is an abridged version of SAAR (Koo et al., 2018) developed by Dr. Hyeongmo Koo. The main difference between the two structures is that the *ESF Tool* uses DotSpatial instead of ArcGIS Engine in SAAR for GIS functions. Hence, the *ESF Tool* does not require a commercial license to run, and it fully supports an open source environment.

Fig. 4.1 illustrates the GUIs of the *ESF Tool*. Fig. 4.1A portrays this software package's main interface with an example map of the (Box–Cox transformed) 2010 population density for Texas counties. The button bar contains basic GIS functions that DotSpatial furnishes. Fig. 4.1B displays the menu items for regression analysis, including linear regression, GLMs,

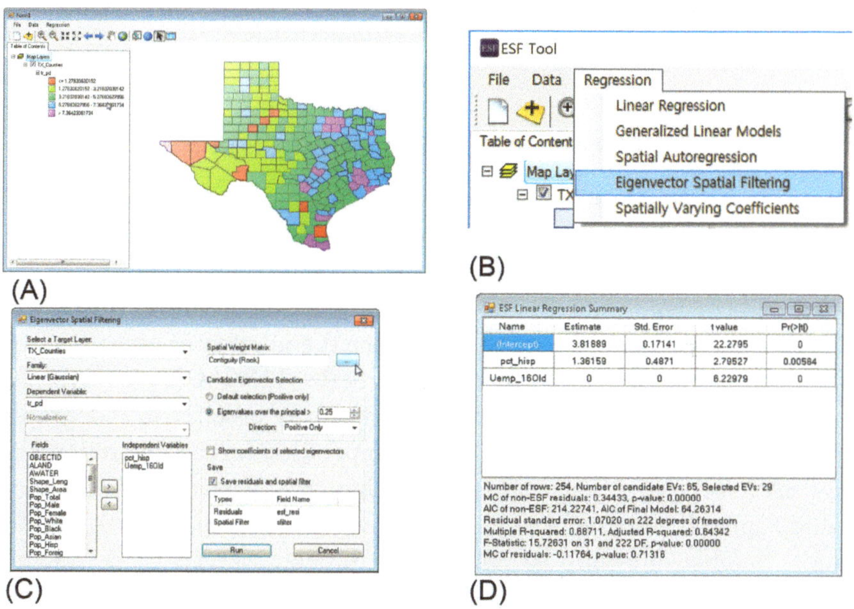

Fig. 4.1 The *ESF Tool* interface (access https://github.com/esftool/esftool for its setup file). (A) The main interface with a map of Box–Cox-transformed 2010 Texas population densities. (B) The menu items for regression functions, including standard regression and MESF models. (C) A GUI for MESF. (D) A GUI displaying MESF results.

[3] Currently only the shapefile of Esri, Inc. is supported.

spatial autoregressive models, and MESF models. Fig. 4.1C presents the pop-up window when the Eigenvector Spatial Filtering item is chosen. With this pop-up window, a user can choose options for a model specification, spatial filter, and residual output to save to the attribute table. In this example, the transformed 2010 population density is selected as a dependent variable, and two variables (the natural logarithm of 2010 median household income and unemployment rates) are selected as covariates. Fig. 4.1D provides results of the MESF specified in Fig. 4.1C. It shows that the MESF successfully accounts for positive SA (PSA) in the residuals of the non-ESF model (i.e., the linear regression model): the MC of the linear regression residuals is 0.48251, with a *p*-value of 0.0000, and the MC of the MESF residuals is −0.11308, with a *p*-value of 0.68202.

Again, MESF functions in the *ESF Tool* are implemented with R internally. Sample R code for implementing MESF can be found in Chun and Griffith (2013) and the accompany website.[4] In addition, other program code, including MATLAB and Fortran, to extract eigenvectors appears in Griffith and Chun (2014).

4.2 Geographic scale and resolution issues for ESFs

Geographic scale refers to the size of a geographic landscape. Geographic resolution refers to the size of the areal units used to partition a geographic landscape. The first concept relates to increasing domain sampling, whereas the second concept relates to infill sampling. Both determine n, the number of areal units included in a study, and hence the number of eigenfunctions that can be extracted from the n-by-n matrix $(\mathbf{I} - \mathbf{11}^T/n)\mathbf{C}(\mathbf{I} - \mathbf{11}^T/n)$. The number of PSA eigenfunctions for a regular square tessellation and a rook's adjacency definition is roughly n/2. The upper limit for the number of negative SA (NSA) eigenfunctions is 3n/4 (Elphick & Wocjan, 2017); a more realistic upper bound in practice is 2n/3 (Taliceo & Griffith, 2018). In other words, roughly one-third to one-half of the eigenvectors for a modified SWM could constitute a candidate set for most georeferenced data analyses. Stepwise selection can be numerically intensive because of the combinatorial number of possible subset regressions; this situation becomes even more challenging as n increases.

Execution time depends, in part, on the amount of random access memory (RAM) and the number of central processing units (CPUs) available for

[4] https://studysites.sagepub.com/chun_griffith/main.htm.

computation, coupled with virtual memory that uses space on a hard drive. In general, once n is in the low thousands, MESF becomes challenged by problem size: stepwise regression must select from 1000 or more candidate eigenvectors, with the possible number of decisions being as many as $2^{n/2}$. Furthermore, when n is large for a remotely sensed image, with hundreds of thousands or millions of pixels, MESF becomes infeasible in practice. Previous experiments suggest that neither sampling of areal units nor aggregation to a coarser resolution adequately solves this problem. The *spmoran* package offers one possible solution here: function *meigen_f* produces fast approximations to the eigenvectors of matrix $(\mathbf{I}-\mathbf{11}^T/n)\mathbf{C}(\mathbf{I}-\mathbf{11}^T/n)$. Pace, LeSage, and Zhu (2013) recommend only including all of a small subset of the eigenfunctions with the largest eigenvalues. Another possible solution is to devise ESF, rather than individual eigenvector, approximations; this strategy is the topic of Section 4.4.

4.3 Determining the candidate set of eigenvectors

Determining an appropriate candidate set of eigenvectors for stepwise regression is a topic addressed in Section 3.2.1. One criterion is to restrict attention only to those eigenfunctions that have the same SA nature as the n calculated regression residuals. This set can be further refined by selecting the subset of eigenfunctions for which $MC_j/MC_{max} \geq 0.25$, where MC_{max} denotes the maximum positive MC value; this threshold value of 0.25 often signifies that an eigenvector holds at least 5% redundant information attributable to SA. Meanwhile, the equation developed by Chun, Griffith, Lee, and Sinha (2016, p. 75; see Section 3.2.1 in Chapter 3) indicates, for the case of PSA, the number of eigenfunctions, beginning with the principal one, to include in a candidate set based upon the degree of SA in regression residuals; this number increases as the degree of SA increases. Finally, the R *SpatialFiltering* function suggests that all eigenfunctions whose eigenvalues deviate from zero by at least a prespecified amount should be considered.

Establishing the wrong candidate set of eigenvectors can present a number of challenges. Including too many can result in overcorrection for SA; including too few can result in undercorrection for SA. Including ones representing the wrong nature of SA can compromise the parsimony of a constructed ESF. This situation is complicated by the existence of mixtures of PSA and NSA, which may well be a primary reason why spatial analysts rarely uncover empirical cases of NSA.

4.4 Extensions to large georeferenced datasets: Implications for big spatial data

Curry (1970, 1971) discusses a cascading approach to spatial data analysis that suggests a practical way to approximate the construction of ESFs for a large or massive spatial tessellation, such as a remotely sensed image. This cascading approach organizes a given spatial tessellation into a multilevel hierarchical structure with multiple segmentations at each level, similar to a quadtree data structure. As an example of a three-level structure, a spatial tessellation is divided into multiple mutually exclusive and collectively exhaustive segments (i.e., subregions) at the first (or finest-resolution) level, with a nested aggregation of these segments making up segmentations at the second level. The third (coarsest-resolution) level tessellation is constructed by aggregating the second-level segmentations. For a regular square tessellation (e.g., a remotely sensed image), this structure is similar to an image pyramid in remote sensing. This cascading approach uses a divide-and-conquer tactic. In other words, it constructs ESFs separately for each segmentation at the finest geographic resolution level, with the means of these fine-resolution analyses constituting the next coarser geographic resolution level; an ESF has a mean of zero, by construction (i.e., each eigenvector in a linear combination has a mean of zero). An ESF can be estimated for this intermediate geographic resolution, again with segment means constituting the coarsest geographic resolution level, once more for which an ESF can be constructed. These three ESFs relate to the notions of MESF local, regional, and global SA components, for example, with a three-level hierarchical structure. One expectation is that for a massive geographic landscape (e.g., $n > 100,000$), the local ESF approximation accounts for virtually all of the latent SA; massive geographic landscapes contain considerable attribute heterogeneity.

4.4.1 A validation demonstration for approximate ESFs

An application based upon differences between two Landsat 8 remotely sensed images (pre- and post-Hurricane Harvey, which made landfall in 2017; Figs. 2.1 and 4.2A) for the Houston, Texas, metropolitan area demonstrates that the proposed cascading technique provides a computationally efficient approach to implement MESF for large or massive spatial tessellations. This validation involves a 100-by-100 subimage in order to compare exact and approximate ESFs (Fig. 4.2B). Construction of the subimage's

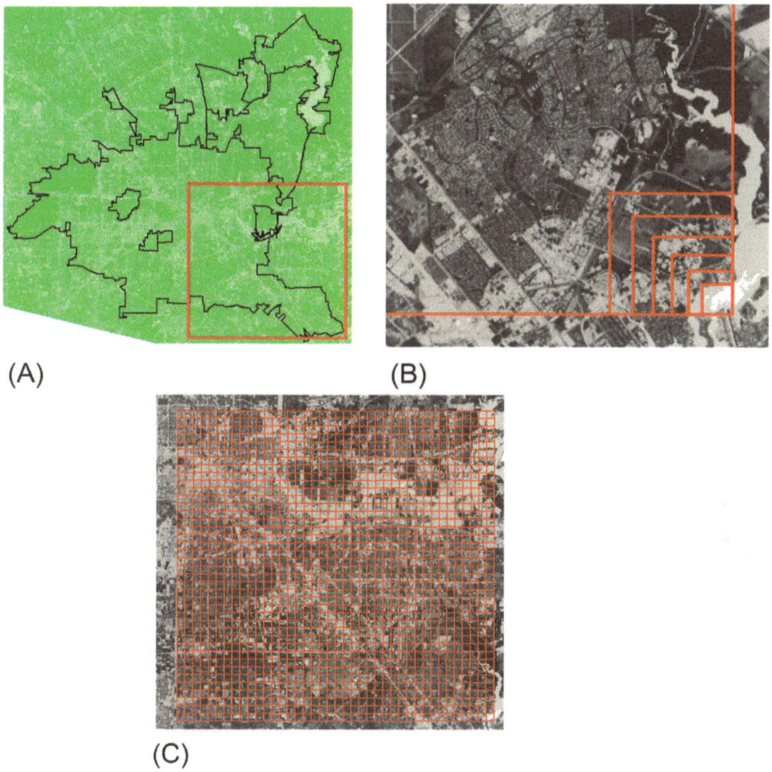

Fig. 4.2 The geographic distribution of the difference between NDVI values for two remotely sensed Landsat 8 images—January 3, 2018, minus April 6, 2017—for part of the Houston, Texas, metropolitan region downloaded from http://glovis.usgs.gov/ (last accessed on April 23, 2018). (A) The location of the 1000-by-1000 specimen subimage (outlined in *red*). (B) A zoom-in of the specimen image with 25-by-25 to 100-by-100 validation subsets demarcated by *red* borders. (C) A superimposed 40-by-40 grid composed of 25-by-25 subgrids for constructing an approximate ESF for the specimen image.

SWM involved the rook's adjacency definition of contiguity (neighborness). The response variable is the normalized difference vegetation index [NDVI $=(B_5-B_4)/(B_5+B_4)$, where B_j denotes spectral band j]; it is the difference between the NDVI values for the pre- and post-Harvey images.

The validation assessment involves employing a portion of the 100-by-100 subimage that expands from its lower right-hand corner, partitioning the selected subimage into mutually exclusive and collectively exhaustive subregions, constructing an ESF for each of these subregions, and then concatenating these ESFs. The first evaluation is for 25 5-by-5 subregions, whereas the sixth evaluation is for 100 10-by-10 sub-regions.[5] The principal trade-off is between the size of the candidate eigenvectors set and the number of linear regressions to be executed.

Fig. 4.3 portrays the overlays of the corresponding exact and approximate ESFs. A close, but not perfect, association exists between these pairs of ESFs. Table 4.1 reveals that strong PSA exists in this subimage geographic

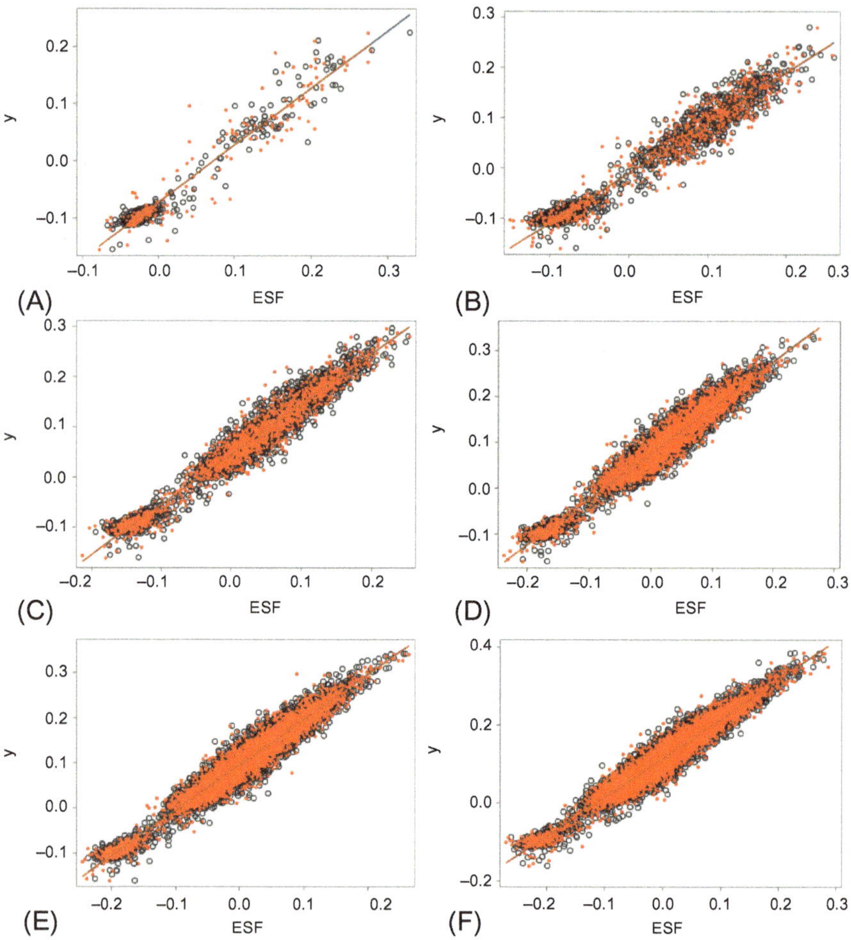

Fig. 4.3 Response variable-ESF scatterplots of approximate ESFs (*red* +) overlaid on exact ESFs (*blue* open circles).

[5] Here the subimage increases in size with an increase in the size of the subregion to preserve exact nesting of the collective set of subregions. This design is similar to increasing domain spatial sampling.

Table 4.1 Summary validation results based on the subimages demarcated in Fig. 4.2B

Size	Subimage size	Subregion size	Number of ESFs	MC	K-S[a]	ESF R^2	MC_e	$\approx$ ESF R^2	$MC_{A\text{-}e}$
5^2-by-5^2	625	5-by-5	25	0.80	0.395	0.95	−0.21 (−0.22)	0.94	0.04 (−0.17)
6^2-by-6^2	1296	6-by-6	36	0.89	0.250	0.96	−0.15 (−0.17)	0.96	−0.03 (−0.17)
7^2-by-7^2	2401	7-by-7	49	0.90	0.149	0.96	−0.16 (−0.17)	0.97	−0.04 (−0.17)
8^2-by-8^2	4096	8-by-8	64	0.87	0.097	0.96	−0.14 (−0.16)	0.96	−0.03 (−0.16)
9^2-by-9^2	6561	9-by-9	81	0.86	0.071	0.95	−0.14 (−0.17)	0.96	−0.03 (−0.16)
10^2-by-10^2	10,000	10-by-10	100	0.86	0.050	0.95	−0.14 (−0.17)	0.96	−0.03 (−0.16)

[a]Results appear to be robust vis-à-vis the normality assumption. In this case, this outcome may be attributable to the bimodal form of the NDVI frequency distribution, which resembles a mixture of two normal distributions (see Appendix 4.A).

Note: K-S denotes the Kolmogorov–Smirnov normality diagnostic statistic; MC_e and $MC_{A\text{-}e}$, respectively, denote exact and approximate ESF residual MCs (expected values are in parentheses).

landscape. It also shows that larger subregions can improve approximate ESF performance. One trade-off is between subregion size and missed border spatial covariations: smaller subregions result in far more border pairings disappearing from the PSA and ESF computations [for a P^2-by-P^2 subimage with P-by-P subregions, this quantity is given by $1/(P+1)$]. This feature could be the reason that trace SA appears to remain in the approximate ESF residuals (i.e., the $MC_{A\text{-}e}$ observed and expected values are not as close as their exact ESF counterparts). Another trade-off is between subregion size and local map pattern component dominance: local map pattern is better captured by smaller subregions. One implication suggested by Table 4.1 is that the subregions' size can be greater than 10-by-10.

4.4.2 An exploration of a massively large remotely sensed image

Fig. 4.2C furnishes an empirical dataset to explore approximate ESFs that cannot be validated directly given prevailing computer technology. Griffith (2015) presents methodology for estimating the spatial autoregressive parameter value for the entire Houston, Texas, differenced images ($n = 4{,}665{,}847$), which is similar to his Fig. 2C (2015, p. 2156). For this Houston NDVI differences image, MC = 0.83 and GR = 0.16.

Table 4.2 summarizes results for a range of subregions. The approximate ESF R^2 values are in keeping with the result reported for other remotely sensed images (e.g., Griffith, 2015). The residual SA should be negative, suggesting that a 5-by-5 subregion is too small. Statistical distribution theory (see Section 3.2.2) for this situation remains to be established. This assessment used a maximum subregion size of 50-by-50 because that scheme already requires 400 assessments of more than 2000 eigenvectors each.

Table 4.2 Summary results based on the 1000-by-1000 pixel specimen geographic landscape and subregions like the ones appearing in Fig. 4.2B

Subregion size	Number of subregions	$\approx$ ESF R^2	Approximate ESF MC	Residual MC
5-by-5	40,000	0.9329	0.8639	0.0282
8-by-8	15,625	0.9498	0.8580	−0.0125
10-by-10	10,000	0.9584	0.8540	−0.0343
20-by-20	2500	0.9758	0.8441	−0.0868
25-by-25	1600	0.9797	0.8413	−0.1008
40-by-40	625	0.9853	0.8371	−0.1218
50-by-50	400	0.9873	0.8354	−0.1310

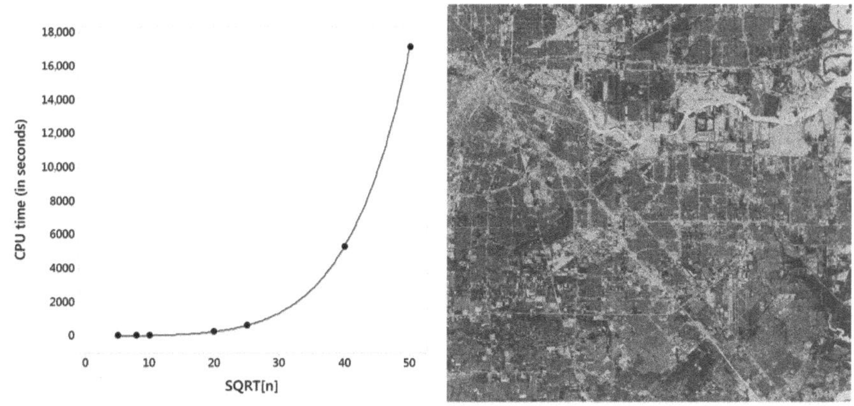

(A) (B)

Fig. 4.4 Output for Fig. 4.2C's 1000-by-1000 pixel geographic landscape. (A) Total CPU time for the set of stepwise regressions. (B) The approximate ESF map based upon a 25-by-25 subregion.

Fig. 4.4A portrays CPU timing results for the selected subregion schemes reported in Table 4.2. The trend line curve is given by

$$\text{Time} = -0.0223 + 0.0223 \times e^{2.7803\left(\sqrt{n_{\text{subregion}}}\right)^{0.4049}}, \; R^2 \approx 1.$$

These results imply that an appealing subregion size is 25-by-25 (more precisely, 25-by-25 to 30-by-30), which involves a candidate set of, at most, 300 eigenvectors for PSA; subregion construction and reading of an external eigenvector file as input add a negligible amount of time to the analysis process (about 6 s of CPU time for this case). Fig. 4.4B corroborates this implication; it portrays the approximate ESF based upon a 25-by-25 subregion, which closely mirrors the original data (Fig. 4.2C).

4.4.3 Correct SWM eigenvectors for a regular square tessellation

One complication associated with regular square tessellation surface partitionings characterizing remotely sensed pixel images arises because reputable computer software packages almost always fail to calculate all of the theoretically correct numerical eigenvectors for the rook adjacency-defined SWM.[6] Being able to detect a problem like this highlights the valuable

[6] The calculated eigenvectors are not wrong. Each pair for a repeated root can be rotated orthogonally any one of an infinite number of ways. Although these vectors are not incorrect, they still almost never are the theoretical ones that result from the appropriate theorem and its proof (Griffith, 2000).

content of Chapter 2. Griffith (2000) derives the correct theoretical eigenvectors for a regular square tessellation forming a complete rectangular region; some are exact, and some are asymptotic for matrix $(\mathbf{I}-\mathbf{11}^T/n)\times \mathbf{C}(\mathbf{I}-\mathbf{11}^T/n)$, whereas all are exact for matrix $\mathbf{C}$. The incorrect numerical eigenvectors complication largely arises from the numerous repeating roots of the eigenvalue polynomial involved (this number increases exponentially with increasing n) and undermines the repeatability and replicability of individual empirical analyses. The two eigenvectors that represent maximum PSA are exact and furnish a good illustrative example of this issue (Fig. 4.5).

The incorrect eigenvectors portraying maximum PSA are, respectively, the sum of (Fig. 4.5C) and the difference between (Fig. 4.5D) the two theoretical eigenvectors, divided by the square root of two; eigenvectors are unique to a multiplicative factor of −1 (to relate properly to their theoretical counterparts, each of these two numerical eigenvectors also needs to be multiplied by −1). Table 4.3 summarizes a regression of each of the incorrect numerical eigenvectors on the pair of theoretical eigenvectors; in all six cases, $R^2=1$. The regression coefficients appearing in Table 4.3 reveal that the different software eigenfunction algorithms also are slightly different.

The impact of these numerical eigenvector deviations from their theoretical counterparts is not pronounced (Table 4.4), as expected; the incorrect eigenvectors merely are orthogonal rotations of their theoretical counterparts. Nevertheless, this impact, although minimal, does compromise the repeatability and replicability of research results; it is beyond rounding error, especially given that these are linear statistical models.

For the recommended 25-by-25 subregion size, the number of eigenvectors with nonzero eigenvalues is 600 (300 representing PSA, and 300

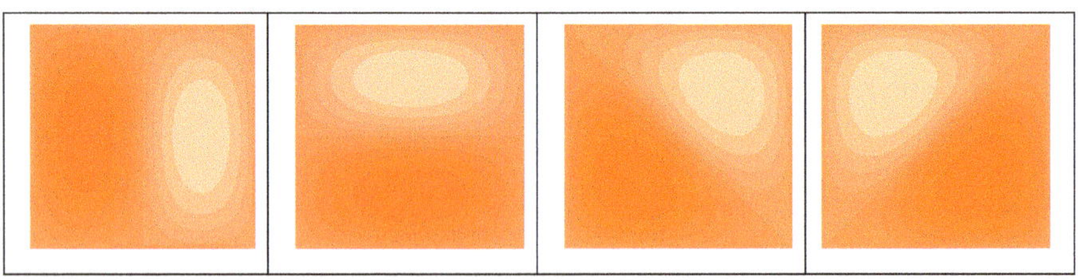

(A) (B) (C) (D)

Fig. 4.5 The first two eigenvectors of matrix $(\mathbf{I}-\mathbf{11}^T/n)\mathbf{C}(\mathbf{I}-\mathbf{11}^T/n)$ for a rook's adjacency definition SWM for a 30-by-30 regular square tessellation, representing maximum PSA; the darkness of *orange* is directly proportional to an eigenvector's element values. (A and B) The correct eigenvectors; and (C and D) examples of common software calculated eigenvectors.

Table 4.3 Regression results for the six duplicate pairs of numerical eigenvectors predicted from their corresponding two theoretical eigenvectors, for a 5-by-5 regular square tessellation and the rook's adjacency definition

		Minitab		R project		SAS 9.4	
Eigenvectors	**Eigenvalue (MC)**	E_h	E_k	E_h	E_k	E_h	E_k
Theoretical $\mathbf{E}_1$	2.7321 (0.8358)	−0.6815	−0.7318	0.6738	−0.7390	−0.6590	−0.7521
Theoretical $\mathbf{E}_2$		−0.7318	0.6815	−0.7390	−0.6738	0.7521	−0.6590
Theoretical $\mathbf{E}_6$	1.0000 (0.3125)	−0.7158	0.6983	0.7237	−0.6901	0.9084	−0.4181
Theoretical $\mathbf{E}_7$		0.6983	0.7158	−0.6901	−0.7237	0.4181	0.9084
Theoretical $\mathbf{E}_8$	0.7321 (0.2288)	−0.8010	−0.5987	0.7619	−0.6477	0.7965	0.6047
Theoretical $\mathbf{E}_9$		−0.5987	0.8010	0.6477	0.7619	0.6047	−0.7965
Theoretical $\mathbf{E}_{16}$	−0.7321 (−0.2288)	0.7676	0.6410	−0.7214	0.6925	0.7683	−0.6401
Theoretical $\mathbf{E}_{17}$		0.6410	−0.7676	0.6925	0.7214	−0.6401	−0.7683
Theoretical $\mathbf{E}_{18}$	−1.0000 (−0.3125)	0.9986	0.0526	−0.7444	−0.6677	−0.9176	−0.3976
Theoretical $\mathbf{E}_{19}$		−0.0526	0.9986	0.6677	−0.7444	−0.3976	0.9176
Theoretical $\mathbf{E}_{23}$	−2.7321 (−0.8358)	0.7068	0.7075	−0.7013	0.7129	0.6787	0.7344
Theoretical $\mathbf{E}_{24}$		−0.7075	0.7068	0.7129	0.7013	−0.7344	0.6787
intercept		0	0	0	0	0	0

Note: Because the eigenvectors are in different orders, h and k denote the corresponding pair for a given duplicate eigenvalue. The theoretical eigenvector numbering is for a descending order of eigenvalues.

Table 4.4 Selected summary results based on a constructed approximate ESF for the 1000-by-1000 pixel specimen geographic landscape using 25-by-25 sub-regions

Eigenvectors source	MC of NDVI	$\approx R^2$	MC_{ESF}	Residual MC
R	0.8236	0.979**6**	0.841**4**	−0.10**10**
SAS	0.8236	0.979**5**	0.841**5**	−0.100**6**
Minitab	0.8236	0.939**6**	0.841**4**	−0.10**10**
Theoretical	0.8236	0.9797	0.8413	−0.1008

Note: Discrepancies are in bold; the theoretical results are correct.

representing NSA). Of these, only 168 numerical eigenvectors are correct; 432 theoretical eigenvectors need to be substituted for their incorrect numerical counterparts. Maintaining this set of eigenvectors in a digital file for access purposes is not resource demanding; for eigenvalues whose elements are calculated to 10 decimal places, the file size is slightly <5 Mb.

4.5 Summary

This chapter examines aspects of implementing MESF. It presents new computer software for implementation summarized by Koo et al. (2018) and demonstrated in Chapter 9. This software dramatically simplifies the employment of MESF, allowing spatial analysts to adopt it with ease. This chapter also discusses other software whose use requires more computer programming skills and is more suitable for advanced spatial analysts who prefer to customize their software analyses. In addition, this chapter introduces an approximation for implementing MESF for massively large georeferenced—in particular, remotely sensed—datasets. In doing so, it authenticates problematic features of, and proposes rectified solutions to, eigenfunction problems for SWMs based upon regular square tessellations and the rook's adjacency criterion. Empirical applications in this chapter are for Texas counties and Landsat 8 images of the Houston region.

Chapters 7 and 8 present additional R code mentioned or alluded to in this chapter. SAAR does not yet handle spatial interaction or space–time data, requiring the creation of dedicated code to implement MESF for these two particular data analytic cases.

Appendix 4.A Frequency distributions of the 10,000 NDVI values for the validation analysis, with superimposed theoretical normal curves

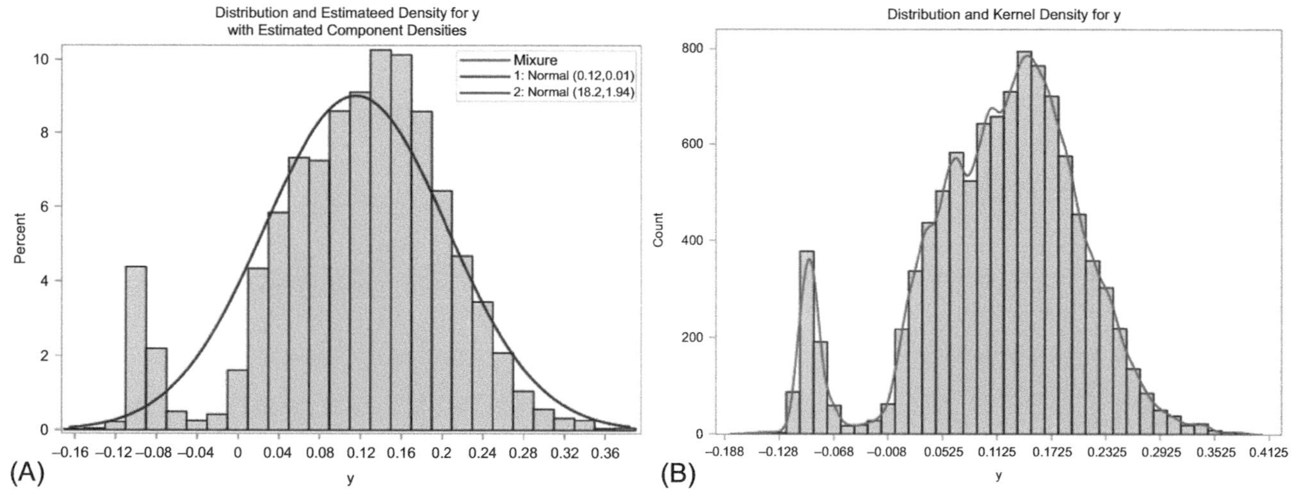

Fig. 4.A1. (A) Single bell-shaped curve superimposed. (B) Kernel density estimation suggesting the presence of two bell-shaped curves.

References

Chun, Y., & Griffith, D. (2013). *Spatial statistics and geostatistics: Theory and applications for geographic information science and technology*. Thousand Oaks, CA: SAGE Publications.

Chun, Y., Griffith, D., Lee, M., & Sinha, P. (2016). Eigenvector selection with stepwise regression techniques to construct eigenvector spatial filters. *Journal of Geographical Systems*, *18*, 67–85.

Curry, L. (1970). Univariate spatial forecasting. *Economic Geography*, *46*, 241–258.

Curry, L. (1971). Applicability of space–time moving-average forecasting. In M. Chisholm, A. Frey, & P. Haggett (Eds.), *Regional forecasting* (pp. 11–24). London: Butterworth.

Elphick, C., & Wocjan, P. (2017). An inertial lower bound for the chromatic number of a graph. *Electronic Journal of Combinatorics*, *25*, #P1.58.

Griffith, D. (2000). Eigenfunction properties and approximations of selected incidence matrices employed in spatial analyses. *Linear Algebra and Its Applications*, *321*, 95–112.

Griffith, D. (2015). Approximation of Gaussian spatial autoregressive models for massive regular square tessellation data. *International Journal of Geographical Information Science*, *29*, 2143–2173.

Griffith, D., & Chun, Y. (2014). Spatial autocorrelation and spatial filtering. In M. Fischer & P. Nijkamp (Eds.), *Handbook of regional science* (pp. 1477–1507). Berlin: Springer-Verlag.

Koo, H., Chun, Y., & Griffith, D. (2018). Integrating spatial data analysis functionalities in a GIS environment: Spatial analysis using ArcGIS Engine and R (SAAR). *Transactions in GIS*, *22*, 721–736.

Murakami, D. 2018. spmoran: Moran's eigenvector-based spatial regression models, v0.1.5. R package version 3.3.3 and higher. https://CRAN.R-project.org/package=spmoran.

Pace, R., LeSage, J., & Zhu, S. (2013). Interpretation and computation of estimates from regression models using spatial filtering. *Spatial Economic Analysis*, *8*, 352–369.

R.NET. (2015). R.NET for users 1.6. Retrieved from http://jmp75.github.io/rdotnet/.

Taliceo, N., & Griffith, D. (2018). *The K_4 graph and the inertia of the adjacency matrix for a connected planar graph*. *183*. (pp. 185–209). STUDIA KPZK PAN (publication of the Polish Academy of Sciences).

CHAPTER 5

MESF and generalized linear regression

Chapter outline

Hundreds of named univariate random variables (RVs) appear in the statistics literature (e.g., Johnson, Kemp, & Kotz, 2005; Johnson, Kotz, & Balakrishnan, 1994, 1995; Leemis & McQueston, 2008). The most common of these RVs employed to represent georeferenced data are the normal (continuous), Poisson (counts), and binomial (percentages). Spatial analysts also consider two special cases of these RVs: the negative binomial as a Poisson RV with a gamma distributed mean and the Bernoulli for presence/absence as a binomial RV with the number of trials, n_{tr}, being one. Besag (1974) describes or alludes to auto-model specifications for these RVs: auto-normal, auto-Poisson, auto-binomial, auto-logistic, and auto-negative binomial (Cressie, 1993, p. 432). Chapter 3 addresses the auto-normal specification. This chapter addresses specifications for the other four types of nonnormal RVs but within the context of Moran eigenvector spatial filtering (MESF) rather than auto-models. The reason for this emphasis is two-fold: (1) neither the auto-Poisson nor the auto-negative binomial can accommodate positive spatial autocorrelation (PSA), which commonly occurs in georeferenced counts data; and (2) none of these other four

https://doi.org/10.1016/B978-0-12-815043-6.00005-7

auto-models has a tractable normalizing constant (a divisor term ensuring that their probabilities sum to one), preventing the use of standard generalized linear model (GLM) estimation methods. This latter complication requires parameter estimation using Markov chain Monte Carlo (MCMC) techniques, which are numerically intensive. In contrast, MESF requires only standard GLM regression routines, dramatically simplifying its estimation; all stepwise regressions in this chapter employ a 10% level of significance for eigenvector selection. Except for the Bernoulli RV, subjecting these GLM RVs to a Box–Cox/Manly transformation may convert them to normal approximations, allowing the use of an auto-normal specification (as in Chapter 3).

Variance homogeneity is a critical property of normal RVs (see Chapter 3). In terms of attribute values, this property refers to variance being constant across all values of Y, resulting in a correlation of zero between Y's mean and variance. One reason to subject Y to a Box–Cox/Manly transformation is to stabilize its variance so that this transformed variance mimics being constant. This property does not hold for the other RVs of interest here. The relationship between the mean and variance for a Poisson RV results in a correlation of one (its mean is identical to its variance). This relationship is weaker for the negative binomial and binomial RVs: $(1+\eta\mu)\mu$ is a function of μ (the mean–variance correlation tends to be >0.968); and, $n_{tr}p(1-p)$ is a function of $n_{tr}p$ (the mean–variance correlation tends to be approximately 0.791). Meanwhile, the Bernoulli RV variance, $p(1-p)$, is a function of two perfectly negatively correlated variables, one being the mean p, yielding a mean–variance correlation of zero. These theoretical relationships establish variance expectations for these RVs. In most georeferenced situations, the observed variances exceed or equal these theoretical variances, with exceedance being called overdispersion. If extra variance characterizes a Poisson RV, then a negative binomial RV may furnish a better description. Conventional treatment of binomial RVs does not involve such a substitution (although a beta-binomial parametric mixture distribution offers a possibility). Finally, with a mean–variance correlation of zero, a Bernoulli RV cannot have overdispersion.

5.1 The logistic regression model specification

Consider a simple dichotomous presence/absence attribute variable, quantified by 1/0 for convenience. The popular treatment of such an indicator variable has been labeled logistic regression because it estimates the probability-defining logistic function $p_i = 1/(1 + e^{\alpha + \mathbf{X}_i \boldsymbol{\beta}_\mathbf{X}})$, $i = 1, 2, \ldots, n$, where α is the intercept, $\mathbf{X}_i$ is a 1-by-P row vector of P covariates for areal unit i, and $\boldsymbol{\beta}_\mathbf{X}$ is a P-by-1 vector of regression coefficients. The following is its MESF logistic regression specification counterpart[1]:

$$\frac{1}{1 + e^{\alpha + \mathbf{X}_i \boldsymbol{\beta}_\mathbf{X} + \mathbf{E}_{K,i} \boldsymbol{\beta}_\mathbf{E}}},$$

where $\mathbf{E}_K \boldsymbol{\beta}_\mathbf{E}$ is the constructed eigenvector spatial filter (ESF), for which $\mathbf{E}_{K,i} \boldsymbol{\beta}_\mathbf{E}$ is its row-vector element for areal unit i, with $\mathbf{E}_{K,i}$ being a 1-by-K row vector of eigenvector elements and $\boldsymbol{\beta}_\mathbf{E}$ a K-by-1 column vector of coefficients for constructing it. For a pure spatial autocorrelation (SA) specification, $\mathbf{X}_i \equiv 0$, $i = 1, 2, \ldots, n$; that is, it disappears from the expression. The SA latent in the geographic distribution of Y can be quantified with join count statistics, which relate to the Moran coefficient (MC) and the Geary ratio (GR) as follows (see Section 1.1.3):

$$\mathrm{MC} = \frac{2n}{\sum_{i=1}^{n} \sum_{j=1}^{n} c_{ij}} \left[\frac{ZZ}{n_z} + \frac{OO}{n - n_z} \right] - 1, \text{and } \mathrm{GR} = \frac{n(n-1)}{\sum_{i=1}^{n} \sum_{j=1}^{n} c_{ij}} \frac{ZO}{n_z (n - n_z)},$$

where n_z denotes the number of areal units with a measure of zero (Z); O denotes a measure of one; and ZZ, OO, and ZO are the standard join count statistics. This MC linkage supports the use of MESF with a Bernoulli RV.

Consider the 77 counties of Texas that are designated as parts of 25 United States (US) Census metropolitan statistical areas (MSAs) in 2010 (Fig. 5.1A). These 77 counties are not distributed randomly across the 254 counties of Texas; rather, they cluster into somewhat dispersed groups. The map-wide probability estimate of a county being part of a MSA, ignoring SA, is 0.3031, whereas the conditional probability taking SA into account is 0.0003; SA accounts for most of this unadjusted probability. A pseudo-R^2 of 0.79 bolsters this result. The ESF represents moderate PSA[2]: MC = 0.65, and GR = 0.40. This ESF contains both a PSA and a negative SA (NSA) component. The PSA component dominates and represents strong SA: MC = 0.72, and GR = 0.32. The NSA[3] component also is strong

[1] Depending upon which group of a dichotomous variable is denoted by zero, this expression may appear as $\frac{1}{1 + e^{-(\mathbf{X}_i \boldsymbol{\beta}_\mathbf{X} + \mathbf{E}_{K,i} \boldsymbol{\beta}_\mathbf{E})}}$.

[2] The maximum MC is 1.098.

[3] The minimum MC is −0.635.

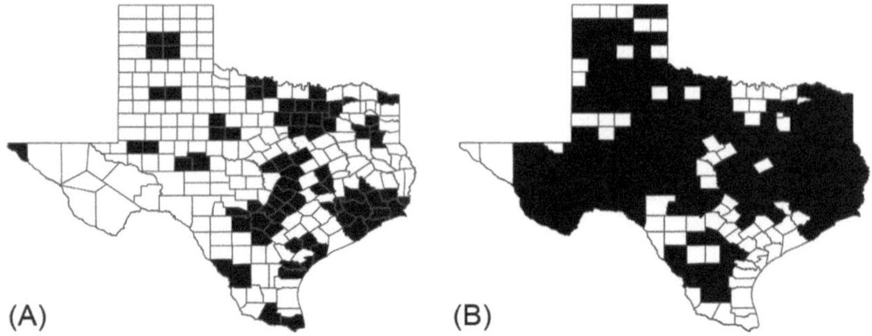

Fig. 5.1 Classified counties of Texas; *black* denotes "yes," and *white* denotes "no" (presence/absence). (A) Designated part of a MSA. (B) Experienced a wildfire by June 30, 2011.

Table 5.1 Cross-classification presence/absence predicted and observed tabulations for the Texas counties

	2010 MSAs		2010 imputed MSAs		2011 wildfires	
Predicted observed	**0**	**1**	**0**	**1**	**0**	**1**
0	**170**	7	**120**	48	**62**	7
1	8	**69**	44	**33**	6	**179**

Note: bold denotes correct classifications.

and substantially interacts with the PSA component: MC = −0.43, and GR = 1.55. The residual SA is modest: MC = −0.10, and GR = 1.02. Classification results, using 0.5 as the threshold, are very good (Table 5.1).

The occurrence of early 2011 wildfires in 185 Texas counties furnishes a second spatial logistic regression example. This example differs from the preceding MSA example because it exhibits fewer clusters—16 versus 25 (Fig. 5.1B)—and hence more continuity. These data[4] are based on the responses from a survey by the Texas Forest Service and local fire departments about roughly 11,000 wildfires across the state, which damaged three million acres and included the destruction of 1339 structures. In this case, the map-wide probability estimate of a county experiencing a wildfire, ignoring SA, is 0.7283, whereas the conditional probability taking SA into account is nearly one; SA accounts for a substantial part of this probability. A pseudo-R^2 of 0.81 bolsters this result. The ESF represents weak to moderate PSA: MC = 0.44, and GR = 0.62. This ESF also contains both a PSA and a NSA component. The PSA component dominates and represents

[4] https://www.texastribune.org/library/data/wildfires-burn-bans-across-texas/.

strong SA: MC = 0.74, and GR = 0.30. The NSA component also is strong and again has a substantial interaction with the PSA component: MC = −0.41, and GR = 1.48. The residual SA is modest: MC = −0.04, and GR = 0.89. Again, classification results, using 0.5 as the threshold, are very good (Table 5.1).

The two illustrations appearing in this section highlight the existence of PSA–NSA mixtures and exemplify the use of MESF as a viable alternative to the auto-logistic specification. Computational results reported here are output from standard GLM procedures and did not require the utilization of MCMC techniques.

5.2 The binomial regression model specification

A binomial RV is an aggregation of Bernoulli RVs. Percentages of, for example, ethnic groups constitute typical georeferenced data frequently conceptualized as a binomial RV (in general, counts of a particular ethnic group divided by counts of the total population). Fig. 5.2A portrays the 2010 geographic distribution of the percentage of Hispanic population across Texas. The 2010 state population was 26,092,033, of which 9,962,643 people self-identified (i.e., were classified in the US Census) as being Hispanic; these quantities yield a statewide value of 38.2%, with individual county percentages ranging from 1% to 96%.

Fig. 5.2A map pattern exhibits a pronounced gradient that decreases from the United States–Mexico border to the northeast Oklahoma–Louisiana

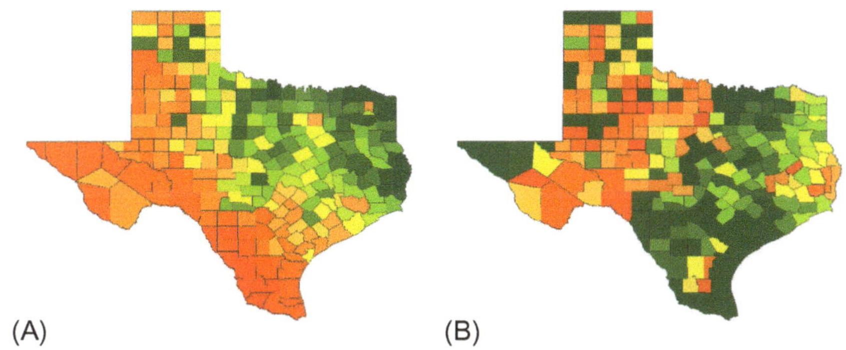

Fig. 5.2 Specimen georeferenced percentage data; the percentage scale is proportional to *dark green* to *yellow* to *dark red*. (A) 2010 Hispanic population. (B) Early 2011 acreage burned by wildfires.

corner of the state. Accordingly, covariates describing this geographic distribution should include county coordinates; the operationalization of these coordinates was with county geometric centroid latitudes and longitudes. The raw 38.2% becomes a conditional 38.4% by accounting for the linear geographic gradient across the state. This conditional percentage decreases to 31.0% by also accounting for latent SA. Meanwhile, the extra binomial variation (i.e., overdispersion) decreases from 20,962 to 6660 by accounting for the linear geographic trend and then to 1319 by accounting for SA. Because total county population ranges from 89 to 4,269,608 and constitutes the individual binomial number of trials, the stepwise eigenvector selection employed a Williams (1982) varying sample size weighting adjustment (i.e., $1/\{1+\gamma[n_i-1]\}$, where γ is estimated with the Pearson deviance statistic[5]).

The linear geographic gradient accounts for roughly 62% of the variation across the state in percentage Hispanic. The PSA ESF, comprising 22 eigenvectors, accounts for roughly 17% of this variation and represents strong PSA (MC = 0.84, and GR = 0.20). The NSA ESF, comprising nine eigenvectors, accounts for roughly 3% of this variance and represents strong NSA (MC = −0.47, and GR = 1.39). The combined ESF represents moderate PSA (MC = 0.53, and GR = 0.48). The residuals (i.e., percentage − predicted percentage) display a small amount of PSA (MC = 0.07, and GR = 0.87).

The preceding example of georeferenced percentage data involves relatively sizable percentages. Fig. 5.2B portrays the early 2011 percentage of acres of land in a county where wildfires burned. These percentages are much smaller, ranging from 0% to 43% of a county's area; slightly <1.5% of all the land acreage in the state burned during the early part of 2011. The overdispersion decreases from 36,752 to 10,798 by accounting for SA. Because the county total acreage ranges from 81,280 to 3,957,568 and constitutes the individual binomial number of trials, quasi-maximum likelihood estimation was employed; this option was exercised because the Williams weights parameter estimation failed to converge during stepwise eigenvector selection.

[5] The quasi-likelihood scaling factor ϕ is estimated by equating the value of Pearson's chi-square statistic for the full model (i.e., all variables are included in the model) to its approximate expected value. Then this estimate of ϕ, $\hat{\phi}$, can be used as a scaling factor (which increases the standard errors of the regression coefficients) during the stepwise selection of eigenvectors, where this weight $[1 + (n_i - 1)\hat{\phi}]^{-1}$ is a function of the denominator (i.e., n_i) of the binomial RV for areal unit i.

The PSA ESF, comprising 26 eigenvectors, accounts for roughly 53% of this variation and represents strong PSA (MC = 0.81, and GR = 0.21). The NSA ESF, comprising six eigenvectors, accounts for roughly 16% of this variance, most of which is an interaction effect with the PSA ESF, and represents strong NSA (MC = −0.40, and GR = 1.35). The combined ESF represents moderate to strong PSA (MC = 0.77, and GR = 0.24). The residuals display trace SA (MC = −0.14, and GR = 1.11).

The two examples appearing in this section illustrate commonly encountered types of percentage data: one with an average percentage close to 50%, and another involving a somewhat rare event, with an average percentage reasonably close to 0% (its results are similar to those for its counterpart with an average percentage reasonably close to 100%). MESF furnishes an efficient and effective methodology to account for SA in these situations.

5.3 The Poisson regression model specification

The Poisson probability mass function affiliated with Poisson regression may be written as follows:

$$P(Y = y_i) = \frac{e^{-\mu_i}(\mu_i)^{y_i}}{y_i!} = \frac{e^{-e^{\left(\alpha + \mathbf{E}_{K,i}\boldsymbol{\beta}_\mathbf{E} + \mathbf{X}_i\boldsymbol{\beta}_\mathbf{X}\right)}}\left[e^{\left(\alpha + \mathbf{E}_{K,i}\boldsymbol{\beta}_\mathbf{E} + \mathbf{X}_i\boldsymbol{\beta}_\mathbf{X}\right)}\right]^{y_i}}{y_i!},$$

where P denotes probability, and y_i is the count for variable Y for areal unit i. The middle and last terms of this equation connect it to regression as follows:

$$\mu_i = e^{\left(\alpha + \mathbf{E}_{K,i}\boldsymbol{\beta}_\mathbf{E} + \mathbf{X}_i\boldsymbol{\beta}_\mathbf{X}\right)} \Rightarrow LN(\mu_i) = \alpha + \mathbf{E}_{K,i}\boldsymbol{\beta}_\mathbf{E} + \mathbf{X}_i\boldsymbol{\beta}_\mathbf{X}$$

where LN denotes natural logarithm. The affiliated regression equation is log-linear, with the term $LN(\mu_i)$ ensuring that μ_i is positive. Replacing the ESF term with an autoregressive term complicates this specification because a multiplicative normalizing constant needs to be determined for it that ensures the infinite sum of these probabilities equals one. This replacement also restricts SA to the NSA range (Besag, 1974) and can only be extended to the PSA range by modifying the sum of the probabilities to be less than one (Kaiser & Cressie, 1997). A MESF specification circumvents these restrictions (Griffith, 2002).

Two features of a phenomenon enable a Poisson RV to characterize it: (1) its values are counts; and (2) it refers to a rare event. This second feature links a Poisson RV to a binomial RV approximation with a very small probability. These features also hold for georeferenced variables. The two examples presented in this section highlight these two features.

5.3.1 Population density

A normal approximation analysis for the 2010 Texas population density appears in Section 3.2 (see Fig. 3.1). The ratios of population counts to county areas were subjected to a Box–Cox power transformation. The numerator of this ratio is a count, which allows it to be analyzed with Poisson regression. In this context, both sides of an equation are multiplied by area, which in turn is converted to log-area and exponentiated in the right-hand side of the equation so that it can become part of the Poisson regression mean response. Because density involves division by area raised to the exponent 1, the Poisson regression coefficient for log-area is set to 1; in other words, it becomes an offset variable.

One statistical diagnostic of importance here assesses $\mu=\sigma^2$. SA tends to induce overdispersion ($\mu<\sigma^2$), as do a number of other aspatial factors: misspecification of the mean function (e.g., missing covariates), nonlinear relationships and covariate interactions, the presence of outliers, heterogeneity in aggregated counts, excessive counts (especially zeros), and positing the wrong probability model. In contrast to the normal approximation results (e.g., Table 3.1), the 2010 Texas population density Poisson regression results produce a pseudo-R^2 of 0.94 (>0.62) and identify 24 (rather than 26) PSA eigenvectors. Meanwhile, the PSA ESF accounts for roughly 81% of the geographic variation in this population density and represents strong PSA (MC=0.78, and GR=0.26). The NSA ESF, comprising 10 eigenvectors, accounts for roughly 13% of this variance, including its moderate interaction effect with the PSA ESF, and represents moderate to strong NSA (MC=−0.37, and GR=1.41). The combined ESF (Fig. 5.3A[6]) represents moderate PSA (MC=0.57, and GR=0.44). The residuals display trace SA (MC=−0.07, and GR=0.87). Finally, accounting for SA reduces overdispersion from 352,987 to 50,677 (quasi-likelihood techniques were employed here); although nearly an 85% decrease occurs in the extra-Poisson variation, the remaining overdispersion still is excessive (50,677 remains very substantially >1).

[6] The urban areas of Amarillo, Austin–San Antonio, Brownsville, Corpus Christi–S. Padre Island, Dallas–Fort Worth, El Paso, Houston–Galveston, Lubbock, and Midland–Odessa are conspicuous in the ESF map.

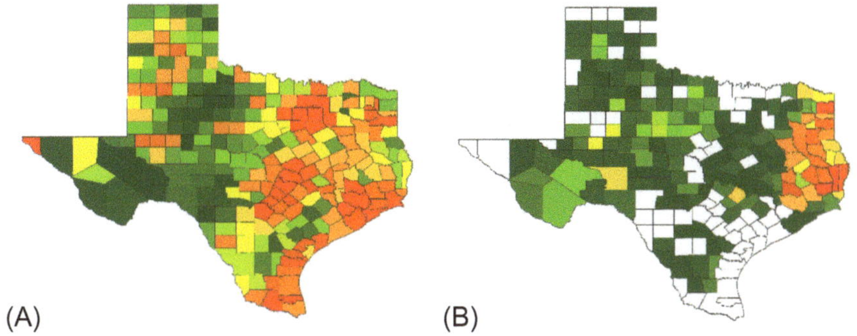

Fig. 5.3 Poisson RV examples for Texas. (A) The 2010 population density ESF geographic distribution across the state from a Poisson regression; ESF values increase from *dark green* to *yellow* to *dark red*. (B) Counts of wildfires; white denotes zero, and the *dark green* to *yellow* to *dark red* scale increases as counts increase from 1 to 97.

5.3.2 Counts of wildfires

Besides attribute variables based upon aggregated counts, such as the population density discussed in the preceding section, geospatial analysts also use Poisson probability models to describe the geographic distribution of point events (i.e., point patterns). In this context, a probability model needs to furnish both a description of the distribution of counts and the locations of the points being counted. The following three important assumptions accompany these analyses: orderliness (i.e., no more than one point can occupy each feasible location); ergodicity [i.e., arithmetic averages of subregions of a single realization of a point pattern over a large geographic landscape (i.e., spatial scale) are the same as point pattern averages for a subregion of the geographic landscape over many realizations]; and, stationarity (i.e., the counts are geographic translation invariant, making them independent of location). The presence of overdispersion often negates some or all of these assumptions.

Fig. 5.3B portrays the counts of numbers of wildfires during early 2011 (unadjusted for county size). More than half of the counties had fewer than three wildfires (69 had zero, 46 had one, and 29 had two). The PSA ESF, comprising 29 eigenvectors, accounts for roughly 75% of the geographic variation in early 2011 counts of wildfires per square mile across Texas and represents very strong PSA (MC = 0.87, and GR = 0.17). The NSA ESF, comprising nine eigenvectors, accounts for roughly 10% of this variance, almost exclusively as an interaction effect with the PSA ESF, and represents strong NSA (MC = −0.42, and GR = 1.43). The combined

ESF represents strong PSA (MC=0.79, and GR=0.24). The residuals display trace SA (MC=−0.06, and GR=0.98). Finally, accounting for SA reduces overdispersion from 16.4 to 2.5 (again, quasi-likelihood techniques were employed here); although nearly an 85% decrease occurs in the extra-Poisson variation, the remaining overdispersion still is excessive (2.5 remains noticeably >1).

5.4 The negative binomial regression model specification

Overdispersion is a complicating factor in each of the preceding two empirical examples. To avoid quasi-likelihood estimation, the solution that frequently comes to mind for addressing an overdispersed Poisson RV is to employ a negative binomial (NB) probability model. The Poisson-gamma parametric mixture formulation often furnishes the conceptual link for this substitution: the mean of a Poisson RV is a gamma RV, introducing additional variation into the distribution of counts. The theoretical expectation is that parameter estimates remain the same, whereas the variance takes a form such as $\mu(1+\eta\mu)$, where η is a dispersion parameter. The gamma mean equals the Poisson mean. One advantage of this specification is that parameter estimation does not require quasi-likelihood techniques.

The NB probability mass function affiliated with NB regression resulting from the preceding Poisson-gamma mixture may be written as follows:

$$P(Y=y_i)=\frac{\Gamma\left(y_i+\frac{1}{\eta}\right)}{\Gamma(y_i+1)\Gamma\left(\frac{1}{\eta}\right)}\left[\frac{\frac{1}{\eta}}{\frac{1}{\eta}+e^{\left(\alpha+\mathbf{E}_{K,i}\boldsymbol{\beta}_E+\mathbf{X}_i\boldsymbol{\beta}_X\right)}}\right]^{\frac{1}{\eta}}\left[\frac{e^{\left(\alpha+\mathbf{E}_{K,i}\boldsymbol{\beta}_E+\mathbf{X}_i\boldsymbol{\beta}_X\right)}}{\frac{1}{\eta}+e^{\left(\alpha+\mathbf{E}_{K,i}\boldsymbol{\beta}_E+\mathbf{X}_i\boldsymbol{\beta}_X\right)}}\right]^{y_i},$$

where Γ denotes the gamma function. Two features of this equation are:

$\dfrac{\frac{1}{\eta}}{\frac{1}{\eta}+e^{\left(\alpha+\mathbf{E}_{K,i}\boldsymbol{\beta}_E+\mathbf{X}_i\boldsymbol{\beta}_X\right)}}$ is of the aforementioned logistic probability form

$\frac{1}{1+e^{LN(\eta)+\alpha+\mathbf{X}_i\boldsymbol{\beta}_X+\mathbf{E}_{K,i}\boldsymbol{\beta}_E}}$, and given the background of binomial regression, the expression $LN(\eta)+\alpha+\mathbf{X}_i\boldsymbol{\beta}_X+\mathbf{E}_{K,i}\boldsymbol{\beta}_E$ defines a regression equation.

As for the Poisson RV case, replacing the ESF term with an autoregressive term complicates this specification because a multiplicative normalizing constant needs to be determined for it that ensures the infinite sum of these probabilities equals one. Again, this replacement also restricts SA to the NSA range, a complication that a MESF specification circumvents.

5.4.1 Population density

A NB description for the 2010 population density across Texas is less parsimonious than its Poisson counterpart (Section 5.3.1): it includes 29 PSA and 14 NSA eigenvectors and accounts for 62% of the variance (a decline from 82%). The PSA ESF still represents strong PSA (MC = 0.80, and GR = 0.22), and the NSA ESF still represents strong NSA (MC = −0.47, and GR = 1.44). The combined ESF still represents moderate to strong PSA (MC = 0.66, and GR = 0.36). The residuals still display trace SA (MC = −0.06, and GR = 1.13). Finally, accounting for SA reduces the overdispersion parameter from 2.4728 to 0.7532 (approximately a 70% reduction in magnitude).

The intercept estimates are roughly the same but not equal: 5.9549 for the Poisson and 6.0387 for the NB, regression, not adjusting for SA; and, 4.7286 for the Poisson and 4.8360 for the NB regression, adjusting for SA. The ESFs are similar but not the same (Fig. 5.4). These results suggest disappointing limitations, implying that perhaps adopting the NB specification here is ill advised. The trade-off is with having to employ quasi-likelihood estimation to handle overdispersion accompanying the Poisson specification.

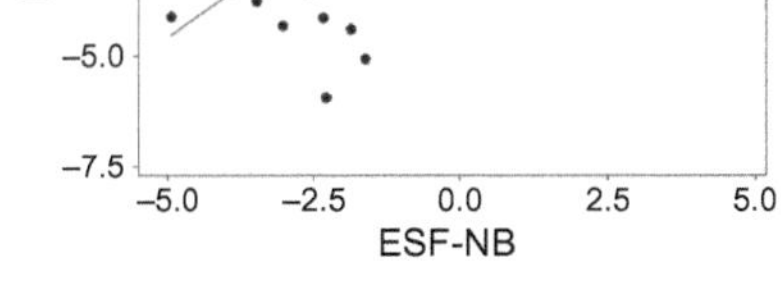

(A)

RV		Poisson	
	Included	yes	no
NB	Yes	E_{1}, E_{2}, E_{3}, E_{4}, E_{11}, E_{13}, E_{15}, E_{18}, E_{20}, E_{23}, E_{25}, E_{28}, E_{29}, E_{35}, E_{38}, E_{39}, E_{43}, E_{52}, E_{56}, E_{61}; E_{167}, E_{174}, E_{190}, E_{212}, E_{248}	E_{7}, E_{8}, E_{9}, E_{17}, E_{21}, E_{26}, E_{27}, E_{30}, E_{53}; E_{173}, E_{218}, E_{220}, E_{223}, E_{235}, E_{237}, E_{241}, E_{245}, E_{254}
	No	E_{9}, E_{31}, E_{47}, E_{54}; E_{177}, E_{186}, E_{187}, E_{200}, E_{214}	

(B)

Fig. 5.4 Poisson–NB comparisons for the 2010 Texas population density. (A) A scatterplot of the Poisson and NB ESFs. (B) Eigenvectors contained in the individual ESFs.

5.4.2 Counts of wildfires

The same stationary, ergodic, and orderly assumptions accompanying a Poisson RV description of point patterns complicate the NB substitution with wildfire counts, too. In terms of point processes, Diggle and Milne (1983) unsuccessfully search for a specification for which the resulting point process satisfies these assumptions while maintaining a NB quadrat count distribution, conjecturing that no such process exists. The Poisson-gamma mixture formulation is a nonergodic and nonstationary point process in which each complete realization is a Poisson RV but whose intensity μ varies between realizations according to a gamma distribution. One issue, like before, is that a negative binomial RV describes the distribution of event counts while not indicating their locations. The beta-negative binomial probability model formulation furnishes an alternative to this specification and constitutes a topic for future research.

Considering the counts of the number of Texas wildfires during early 2011 once more, we assume here that, although a NB fails to furnish an adequate description of many aggregated point count map patterns, we can use a NB to approximate a description of the map pattern of county-aggregated wildfire counts across Texas.[7] Perhaps not surprising, an NB specification displays poor properties. The regression residuals display trace SA ($MC=-0.00$, and $GR=1.27$), suggesting that perhaps the candidate eigenvector set needs expanded. The ESF contains 39 PSA and 10 NSA eigenvectors. This ESF represents strong SA ($MC=0.79$, and $GR=0.24$) and accounts for roughly 56% of the variation in density of wildfire counts across Texas in early 2011. The PSA ESF ($MC=0.87$, and $GR=0.18$) accounts for virtually all of the geographic variation, whereas the NSA ESF ($MC=-0.44$, and $GR=1.48$) principally reduces the magnitude of the dispersion parameter estimate. These disappointing results imply that a NB specification most likely is inadequate here, a contention further supported by Fig. 5.5.

5.5 The selection of eigenvectors to construct an ESF for GLMs

Section 3.2.1 summarizes eigenvector selection for the normal probability model case. The GLM case is similar. The logistic, binomial, Poisson,

[7] Wildfires do not behave like, say, trees, and hence their clustering may not benefit from, for example, a Neyman–Scott clustering specification (Tanaka, Ogata, & Stoyan, 2008).

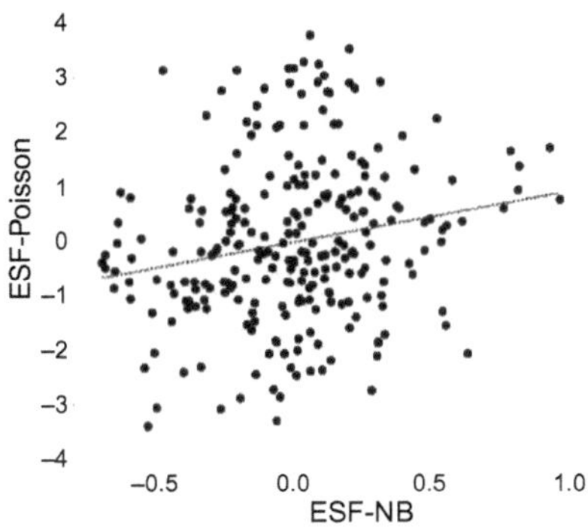

(A)

RV		Poisson	
	Included	yes	no
NB	Yes	E_1, E_2, E_4, E_5, E_6, E_7, E_8, E_{11}, E_{12}, E_{15}, E_{17}, E_{20}, E_{21}, E_{22}, E_{25}, E_{29}, E_{30}, E_{31}, E_{32}, E_{33}, E_{34}, E_{38}, E_{39}, E_{46}, E_{48}, E_{51}; E_{167}, E_{198}, E_{207}, E_{222}, E_{235}, E_{242}, E_{253}	E_3, E_{16}, E_{17}, E_{18}, E_{26}, E_{27}, E_{35}, E_{40}, E_{50}, E_{52}, E_{55}, E_{58}, E_{59}, E_{63}; E_{189}, E_{197}, E_{230}
	No	E_{44}, E_{54}, E_{56}; E_{178}, E_{241}	

(B)

Fig. 5.5 Poisson–NB comparisons for early 2011 Texas wildfires density. (A) A scatterplot of the Poisson and NB ESFs. (B) Eigenvectors contained in the individual ESFs.

and NB regression cases discussed in this chapter can be characterized as log-linear model specifications:

$$\text{logistic}: \mathrm{E}\left[\mathrm{LN}\left\langle\frac{1-\mathrm{p}}{\mathrm{p}}\right\rangle\right] = \alpha\mathbf{1} + \mathbf{X}\boldsymbol{\beta}_{\mathbf{X}} + \mathbf{E}_{\mathrm{K}}\boldsymbol{\beta}_{\mathbf{E}}$$

$$\text{Poisson}: \mathrm{E}[\mathrm{LN}(\boldsymbol{\mu})] = \alpha\mathbf{1} + \mathbf{X}\boldsymbol{\beta}_{\mathbf{X}} + \mathbf{E}_{\mathrm{K}}\boldsymbol{\beta}_{\mathbf{E}}$$

$$\text{NB}: \text{same as Poisson, coupled with } \mathrm{VAR}(\mathrm{Y}) = \mu(1 + \eta\mu).$$

where $\langle\ \rangle$ denotes a column vector, **1** is an n-by-1 vector of ones, and VAR denotes variance. Joint discussion of the Poisson and NB frequently highlight that their regression parameter estimates are the same but that their standard errors differ.

These log-linear forms suggest that normal approximations can be used to determine the candidate set of eigenvectors employing the formula reported in Chun et al. (2016, p. 75) and presented in Section 3.2.1. Maximum likelihood estimation of the GLM regression parameters is equivalent to weighted least squares estimation when probability model assumptions hold. Attaching weights during estimation tends to compromise and corrupt the orthogonality and uncorrelatedness properties of the MESF eigenvectors, although not to an excessive degree. Nevertheless, stepwise procedures can be utilized here, as can the significance level criterion, to construct ESFs.

5.6 ESF prediction with generalized linear regression

Griffith (2010) shows that the expectation–maximization (E-M) imputation is equivalent to the prediction of new observed values. This equivalency dramatically simplifies equations such as Eq. (3.8) in Section 3.4. Rather than estimating a missing response value as a parameter (i.e., a conditional mean), it can be included as a missing value and then predicted from the covariates. In a pure SA case, the redundant information used for prediction is that attributable to SA (and represented by Moran eigenvectors).

Table 5.2 summarizes the leave-one-out imputation for the Texas MSA logistic regression (i.e., Bernoulli RV) example (Section 5.1). Fourteen of the eigenvectors in the complete data ESF are selected for each of the

Table 5.2 Individual eigenvector frequencies in the 254 constructed imputation ESFs

Frequency of occurrence	Eigenvectors
254	E_{2}, E_{3}, E_{4}, E_{9}, E_{23}, E_{25}, E_{28}, E_{29}, E_{35}, E_{38}, E_{44}, E_{53}, E_{56}, E_{61}
243–253 (at least 90%)	E_{11}, E_{24}, E_{46}, E_{57}, E_{190}, E_{200}, E_{221}, E_{225}
209–214 (80%–89%)	E_{5}, E_{20}, E_{42}, E_{226}
184–202 (70%–79%)	E_{10}, E_{13}, E_{54}, E_{59}, E_{168}
157–169 (60%–69%)	E_{1}, E_{30}, E_{52}, E_{241}, E_{244}
133–139 (50%–59%)	E_{18}, E_{205}, E_{227}, E_{248}
103–118 (40%–49%)	E_{180}, E_{213}, E_{219}, E_{252}
26–116 (10%–39%)	E_{21}, E_{32}, E_{33}, E_{37}, E_{48}, E_{172}, E_{186}, E_{213}, E_{214}, E_{215}, E_{219}, E_{235}, E_{236}, E_{247}, E_{252}
1–25 (<10%)	E_{6}, E_{7}, E_{8}, E_{12}, E_{14}, E_{15}, E_{16}, E_{19}, E_{22}, E_{26}, E_{31}, E_{34}, E_{39}, E_{40}, E_{41}, E_{45}, E_{47}, E_{49}, E_{50}, E_{51}, E_{55}, E_{58}, E_{60}, E_{62}, E_{63}, E_{64}, E_{65}, E_{166}, E_{167}, E_{169}, E_{171}, E_{174}, E_{175}, E_{176}, E_{177}, E_{178}, E_{179}, E_{181}, E_{182}, E_{183}, E_{184}, E_{187}, E_{188}, E_{189}, E_{192}, E_{193}, E_{194}, E_{195}, E_{196}, E_{197}, E_{198}, E_{199}, E_{201}, E_{202}, E_{203}, E_{206}, E_{207}, E_{209}, E_{211}, E_{212}, E_{216}, E_{217}, E_{218}, E_{220}, E_{222}, E_{224}, E_{228}, E_{230}, E_{231}, E_{233}, E_{234}, E_{238}, E_{239}, E_{240}, E_{242}, E_{246}, E_{249}, E_{250}, E_{251}, E_{254}

Note: Italic denotes the eigenvectors included in the constructed complete data ESF.

254 imputation ESFs; all but one are included in at least half of these imputation ESFs. Only two eigenvectors are somewhat problematic ($\mathbf{E}_{227}$ and $\mathbf{E}_{219}$). Eighty eigenvectors enter into various imputation ESFs as noise (i.e., <10% of these constructions). The correlation (i.e., phi coefficient) for the imputed and observed values is 0.155; Table 5.1 presents the accompanying classification table.

Fig. 5.6 portrays observed-versus-imputed value scatterplots. Fig. 5.6A is for the percentage of Hispanics, was estimated with quasi-likelihood techniques—the overdispersion weight of $1/[1+0.046835(\text{population}-1)]$ was calculated with the complete dataset—and has a correlation of 0.87. This analysis also produces results like those appearing in Table 5.2. Fig. 5.6B is for population density as a Poisson RV. The stepwise regression for it was executed with quasi-likelihood estimation techniques

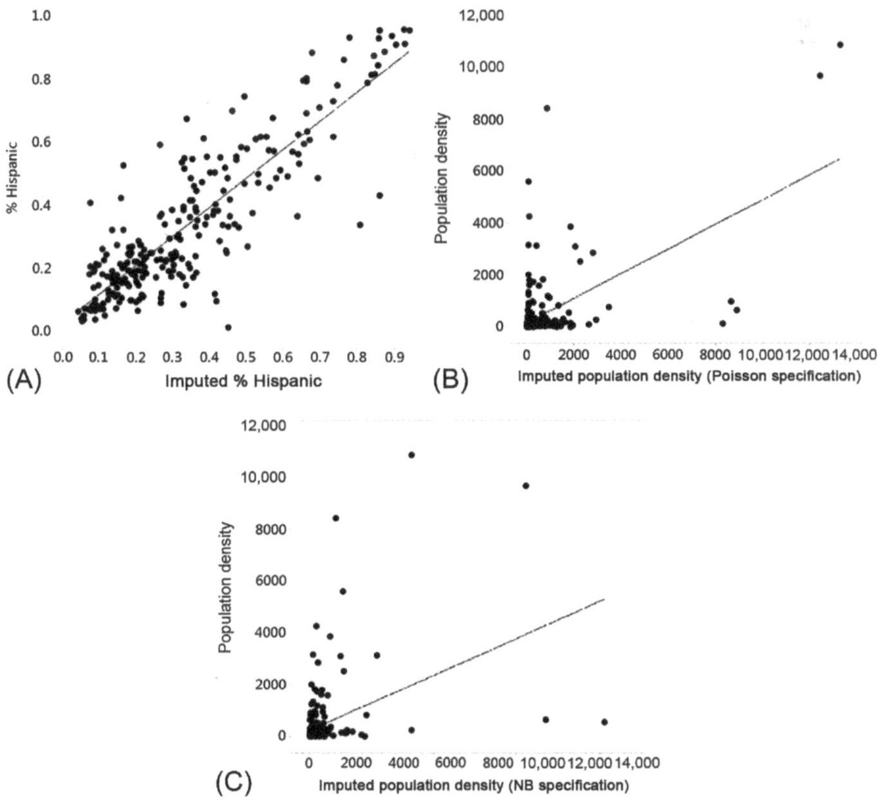

Fig. 5.6 Scatterplots of observed values versus ESF imputations. (A) Percentage of Hispanic population. (B) Population density, a Poisson specification. (C) Population density, a NB specification.

Table 5.3 The PRESS/ESS index for assessment of ESFs constructed for selected RVs

	Bernoulli	Binomial	Poisson	NB
$PRESS_F/ESS$	3.270	1.408	1.001	1.001
$PRESS_V/ESS$	7.957	1.681	1.002	1.397

Note: subscript F denotes a fixed construction, and subscript V denotes a variable construction.

(scale parameter=278). It has a correlation of 0.59. This analysis also produces results like those appearing in Table 5.2. In addition, Fig. 5.6C is for population density but as a NB RV. Similar to the binomial RV analysis, its analysis involved a new stepwise selection of eigenvectors for each imputation. It also produces results like those appearing in Table 5.2 and has a correlation of 0.41.

The ratio of the predicted error sum of squares (PRESS) and the error sum of squares (ESS) furnishes an informative specification assessment index. This index may be calculated in two different ways: setting the ESF to that constructed with the complete dataset [a fixed (F) construction], and constructing a new ESF for each of the n imputations [a variable (V) construction]. The latter is the context in which Table 5.2 and Fig. 5.6 were created. Table 5.3 summarizes these two indices for the imputation exercises presented in this section. Because it seeks to furnish a better fit to the $(n-1)$ observed values in a cross-validation, the variable construction should exhibit a poorer performance than the fixed construction (Table 5.3). But this variable construction would be the one employed in practice. The percentage Hispanic ESF displays the best performance here. Both the Poisson and the NB produce cross-validation types of results that are equivalent to their corresponding complete data model fits. Fig. 5.6B and C, however, underscores that they are not precisely describing the population density data.

5.7 Summary

This chapter presents a theoretical foundation for MESF in the context of nonnormal RVs (i.e., Bernoulli, binomial, Poisson, and NB, with both small and sizable global probability parameters) and classical GLM regression. It highlights problems with and solutions for the presence of overdispersion in georeferenced data, relating this extra variance to map pattern and SA. It demonstrates how to construct an ESF using GLM regression and either quasi- or regular maximum likelihood estimation, and then how to evaluate its quality with selected diagnostics. The primary approach presented for ESF construction is a stepwise selection of regression covariates procedure (employing a 10% level of significance selection criterion); implementing this strategy requires treating eigenvectors as covariates. Once constructed, an ESF needs to be interpreted; this interpretation is based upon its MC and GR indices, the statistical distribution it mimics, and its ability to replicate the map pattern displayed by its corresponding response variable Y. One of the biggest advantages of MESF here is the avoidance of auto-models requiring MCMC techniques for estimation purposes and circumvention of constraints such as the Poisson and the NB RVs being unable to accommodate PSA. Finally, MESF supports predication in the tradition of imputations. Overall, this chapter establishes the fundamental foundation for implementing MESF for nonnormal RVs, building on the existing large body of GLM theory.

References

Besag, J. (1974). Spatial interaction and the statistical analysis of lattice systems. *Journal of the Royal Statistical Society B, 36*, 192–225.

Chun, Y., Griffith, D., Lee, M., & Sinha, P. (2016). Eigenvector selection with stepwise regression techniques to construct eigenvector spatial filters. *Journal of Geographical Systems, 18*, 67–85.

Cressie, N. (1993). *Statistics for spatial data* (Revised ed). New York: Wiley.

Diggle, P., & Milne, R. (1983). Negative binomial quadrat counts and point processes. *Scandinavian Journal of Statistics, 10*, 257–267.

Griffith, D. (2002). A spatial filtering specification for the auto-Poisson model. *Statistics and Probability Letters, 58*, 245–251.

Griffith, D. 2010. Some simplifications for the expectation-maximization (EM) algorithm: The linear regression model case. InterStat, March article 2 (http://interstat.statjournals.net/YEAR/2010/abstracts/1003002.php, http://interstat.statjournals.net/), 23 pp.

Johnson, N., Kotz, S., & Balakrishnan, N. (1994). *Continuous univariate distributions. vol. 1* (2nd ed.). New York: Wiley.

Johnson, N., Kotz, S., & Balakrishnan, N. (1995). *Continuous univariate distributions. vol. 2* (2nd ed). New York: Wiley.

Johnson, N., Kemp, A., & Kotz, S. (2005). *Univariate discrete distributions* (3rd ed.). NY: Wiley.

Kaiser, M., & Cressie, N. (1997). Modeling Poisson variables with positive spatial dependence. *Statistics and Probability Letters, 35*, 423–432.

Leemis, L., and J. McQueston. 2008. Univariate distribution relationships. American Statistician 62: 45–53. http://www.math.wm.edu/~leemis/chart/UDR/UDR.html.

Tanaka, U., Ogata, Y., & Stoyan, D. (2008). Parameter estimation and model selection for Neyman-Scott point processes. *Biometrical Journal, 50*, 43–57.

Williams, D. (1982). Extra-binomial variation in logistic linear models. *Applied Statistics, 31*, 144–148.

CHAPTER 6

Modeling spatial heterogeneity with MESF

Chapter outline

In a statistics context, heterogeneity (an antonym of homogeneity) relates to the identically distributed part of the independent and identically distributed (IID) assumption (see Section 1.2) attached to a dataset for data analytic convenience purposes (e.g., simplification of mathematical statistics algebra): this concept means that statistical features (especially means, variances, and/or frequency distributions) vary over subsets of a dataset. Section 5.3.2 alludes to this notion in its discussion of ergodicity [i.e., every sequence or sizable sample is equally representative of the entire geographic landscape with regard to, say, parameters such as the mean, variance, and spatial autocorrelation (SA)], and stationarity (i.e., the stochastic/random process under study has a constant unconditional[1] joint probability distribution, and hence

[1] In this context, the unconditional adjective refers to independent observations (i.e., no SA). Besag (1974) shows how messy the probability distribution becomes when this feature does not hold; the need to employ Markov chain Monte Carlo (MCMC) techniques to simulate nonnormal auto-random variables shows how complicated simulating a conditional joint probability distribution can be.

Spatial Regression Analysis Using Eigenvector Spatial Filtering
https://doi.org/10.1016/B978-0-12-815043-6.00006-9

constant parameters, across a geographic landscape[2]; this property allows the inverse of a spatial linear operator to have its matrix expansion converge—see Section 3.1 and Eq. (3.2). The relevance of the georeferenced data heterogeneity issue is of two forms: (1) attribute and (2) geographic. This relevance alludes to the important fundamental question of whether data heterogeneity matters. In the attribute domain, data heterogeneity results in greater variance than its homogeneous counterpart, which motivated the invention of Box–Cox/Manly transformations as well as generalized linear model (GLM) theory, among other data science developments (e.g., weighted least squares). In the spatial domain, heterogeneity signifies the presence of local, rather than global, data relationships (variable associations whose correlations with a response variable vary across a geographic landscape), including spatial dependencies [e.g., local indices of SA, such as local indicators of spatial association (LISA) and the Getis–Ord G_i statistic]. Conventional statistical methodology sometimes handles heterogeneity by allowing mean response and/or variance parameters to vary across an attribute's magnitude and/or across a geographic landscape. This chapter addresses Moran eigenvector spatial filtering (MESF) approaches to account for heterogeneity by allowing impacts of SA to vary across space. It extends the eigenvector spatial filter (ESF) formulation presented in preceding chapters that accounts for heterogeneity by allowing only a regression intercept term to vary across space (Section 3.1.2). It relates more directly to the simultaneous autoregressive (SAR) than to the autoregressive response (AR) model conceptualization.

6.1 Spatially varying coefficients

Spatially varying coefficients (SVCs) may be defined as nonconstant regression coefficients that vary by areal unit across a geographic landscape in order to account for nonstationarity in a relationship between a response variable and a given covariate, rendering a statistical explanation of data heterogeneity; they allude to the notion of local regression. Mixed statistical models furnish one conceptualization for SVCs (see Waller, Zhu, Gotway, Gorman, & Gruenewald, 2007), a topic discussed in Section 3.1, because their model specifications introduce a random effects (RE) term into a probability model

[2] In the case of a Gaussian geographic random variable, stationarity means that the underlying probability distribution at each location is the same normal distribution, which has the same mean and the same variance.

to capture SA in a way that it influences estimated associations. The linear regression model equation capturing this conceptualization is

$$\mathbf{Y} = \mathbf{X}\boldsymbol{\beta}_{\mathbf{X}} + \mathbf{E}_K\boldsymbol{\beta}_{\mathbf{E}} + \mathbf{Z}_{SURE} + \boldsymbol{\varepsilon} = \mathbf{X}\boldsymbol{\beta}_{\mathbf{X}} + \boldsymbol{\xi}, \tag{6.1}$$

where $\boldsymbol{\xi}$ is an n-by-1 regression residual error term, $\boldsymbol{\varepsilon}$ is an n-by-1 $N(\mathbf{0}, \mathbf{I}\sigma^2)$ standard IID error term, the ESF $\mathbf{E}_K\boldsymbol{\beta}_{\mathbf{E}}$ is a spatially structured (i.e., accounts for SA) random effects (SSRE), and $\mathbf{Z}_{SURE}$ is a spatially unstructured (i.e., void of SA) random effects (SURE). Estimating the RE term $(\mathbf{E}_K\boldsymbol{\beta}_{\mathbf{E}} + \mathbf{Z}_{SURE})$ and separating it from the residual error $\boldsymbol{\varepsilon}$ requires additional information, such as repeated measures furnished by a space–time series or prior distributions posited in a Bayesian analysis (e.g., Griffith, 2013). The addition of a RE term (SSRE + SURE) to a spatial model specification allows a borrowing of information across locations: in its simplest form, as only a spatially varying intercept term, data from all locations inform about the global intercept, whereas data from each individual location inform about each location's deviation from this global intercept. In other words, $(\mathbf{1}\beta_0 + \mathbf{E}_K\boldsymbol{\beta}_{\mathbf{E}} + \mathbf{Z}_{SURE})$, where $\mathbf{1}$ denotes an n-by-1 vector of ones, is centered on the global intercept β_0, with deviations from this constant defined by $(\mathbf{E}_K\boldsymbol{\beta}_{\mathbf{E}} + \mathbf{Z}_{SURE})$. Georeferenced data do not need to be independent to define inference in this context. The SSRE component captures SA latent in the geographic distribution of Y attribute values, allowing its inferential and estimation basis to be the broad mathematical statistical theory underlying generalized linear mixed models (i.e., GLMs with random effects). The common assumption here is that the RE term is normally distributed, with a mean of zero [the regression residual is $(\mathbf{E}_K\boldsymbol{\beta}_{\mathbf{E}} + \mathbf{Z}_{SURE} + \boldsymbol{\varepsilon})$], and uncorrelated with both the covariates (matrix $\mathbf{X}$) and the residuals.

A RE term requires additional information to separate it from stochastic error $\boldsymbol{\varepsilon}$ that together constitute a residual $\boldsymbol{\xi}$. As the preceding discussion emphasizes, the SSRE component arises from SA and can be captured by an ESF. It exploits the SA latent in regression residuals, correlation that constitutes the additional information needed to identify it. Additional information also is a prerequisite for identifying a SURE component. To date, this supplementary information has been of the following two kinds: (1) repeated measures for observations and (2) Bayesian priors. Griffith (2013) presents evidence that these two disparate sources render essentially the same RE estimates and that, as a form of repeated measures, the length of a time series and the spacing of measurements in time (within reason) have relatively little effect on RE estimates.

Discussion about how to estimate the SURE component appears in Chapter 8 of this book in the context of space–time data series analysis; Bayesian techniques are beyond the scope of this book.

Population density values calculated with the annual Texas county inhabitants estimates for the time period 2010–16[3] constitute a 254-by-7 space–time data series. The calculation of these estimated counts of people is derived from the demographic balancing equation, which begins with a base number of residents (the previous year's population estimate, which is validated every 10 years by a national decennial census), adds the number of births, subtracts the number of deaths, and adds both the domestic and international net migration counts (based on official public records). The 2010 ACS population density (see Sections 5.3.1 and 5.4.1) exhibits considerable overdispersion and moderate to strong positive SA (PSA). Results based on annual estimates of population density calculations that appear in Table 6.1 are consistent with this earlier finding. They imply considerable heterogeneity across space that increases through time. Not only because of its recognized weaknesses but also because of its poor description of 2010 Texas population density, the negative binomial (NB) depiction presented in the preceding chapter has been deemed inadequate. A spatially varying intercept specification offers an appealing alternative to it. The RE term

Table 6.1 Estimation results pertaining to random effects for a space–time dataset

	Constant intercept		Spatially varying intercept				
Year	$\hat{\beta}_0$	Scale	$\hat{\beta}_0$	$\hat{\beta}_{RE}$	Scale	Pseudo-R^2	Residual variance
2010	5.9219	581.6276	4.3991	0.9906	7.9734	0.9995	26.3^2
2011	5.9377	587.5374	4.4048	0.9937	6.0547	0.9998	19.5^2
2012	5.9541	593.9798	4.4091	0.9974	3.9428	0.9999	13.3^2
2013	5.9694	599.8358	4.4144	1.0005	1.4835	1.0000	3.8^2
2014	5.9870	606.4307	4.4224	1.0034	2.8768	1.0000	9.3^2
2015	6.0049	612.9132	4.4325	1.0058	6.0648	0.9998	20.6^2
2016	6.0205	618.5099	4.4426	1.0075	9.2028	0.9994	32.1^2

[3] Estimated population counts were retrieved from https://www.tsl.texas.gov/ref/abouttx/popcnty2010-11.html and represent population estimates as of July 1 of each year. US Census 2010 population counts were retrieved from https://www.tsl.texas.gov/ref/abouttx/popcnty12010.html. These web pages were last accessed on March 6, 2018. These two sets of 2010 counts differ from the American Community Survey (ACS) counts discussed in the Preface of this book.

has a mean near zero (0.0007), closely conforms to a bell-shaped curve (a Shapiro–Wilk diagnostic statistic probability of 0.71), accounts for an enormous amount of the extra-Poisson variation (Table 6.1), and furnishes an extremely good description of the population densities, with the identified residual term accounting for less than about 1% of the population density variance across Texas. As anticipated, the RE term furnishes its best description for the central part of the space–time data series. Furthermore, inclusion of the RE term results in intercept estimates that are much closer to the raw population density means; they are about a third smaller than their intercept-only counterpart estimates.

The RE term can be partitioned into a SSRE and SURE component (Fig. 6.1); the former component accounts for roughly 65% of its variance, whereas the latter component accounts for about 35% of its variance. The SSRE component is an ESF constructed with 24 PSA and five negative SA (NSA) eigenvectors (Fig. 6.1A); its SA is strong (MC = 0.81; GR = 0.24). Of these 29 eigenvectors, 18 PSA and two NSA vectors appear in the previously constructed ESF (Fig. 5.4A). In addition, this space–time-based ESF contains three PSA and two NSA eigenvectors not appearing in the Poisson ESF, but appearing in the NB ESF (perhaps relating to extra-Poisson variation). It also contains four (three PSA and one NSA) eigenvectors that appear in neither of these previous Poisson or NB ESFs. Patuelli, Griffith, Tiefelsdorf, and Nijkamp (2011) address the stability of ESF composition over time and suggest that these types of nuances are to be expected. Furthermore, this space–time-based ESF closely conforms to a bell-shaped curve (a Shapiro–Wilk diagnostic statistic probability of 0.84). The conspicuous map pattern in Fig. 6.1A appears in terms of the the three major

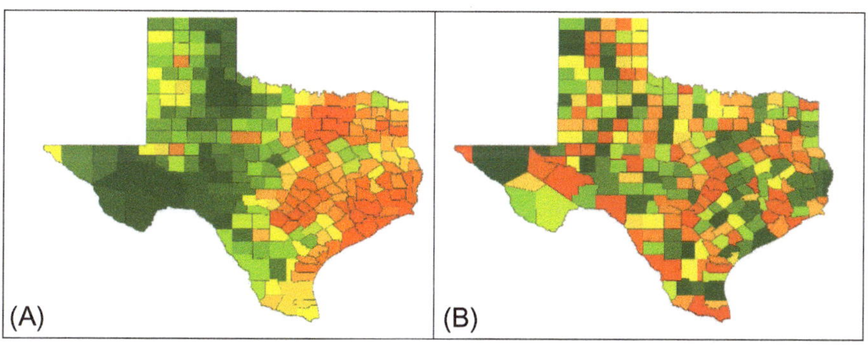

Fig. 6.1 The 2010–16 Texas population density RE term. (A) Its SSRE component. (B) Its SURE component.

metropolitan regions of Texas: Dallas–Fort Worth, Austin–San Antonio, and Houston (see Fig. 5.3).

Fig. 6.1B portrays the SURE; it contains a trace amount of SA (MC = −0.00; GR = 0.91) and deviates somewhat from a bell-shaped curve (a Shapiro–Wilk diagnostic statistic of 0.979, with a probability of 0.001). The SSRE accounts for roughly 60% of the geographic variation in population density each year, whereas the SURE accounts for roughly 12%, with another 28% attributable to some type of interaction effect when these two terms jointly appear in a nonlinear Poisson mean specification.

6.2 An ESF expansion of regression coefficients

The MESF specifications appearing in earlier chapters and the preceding section are for a geographically varying intercept term only (see Section 3.1.2), which might be incomplete for a single time period because the SURE component is not separated from the residuals, alluding to it furnishing a second conceptualization for spatially varying regression coefficients. This particular extended specification relates to the SAR (spatial error) model (Tiefelsdorf & Griffith, 2007) and supports geographically varying regression slope coefficients in the tradition of geographically weighted regression (GWR). Two traditional statistical concepts are necessary here. The first, highlighted by analysis of covariance (ANCOVA), includes 0–1 indicator variables in a data analysis to differentiate between subgroups of observations; these subgroups are the source of data heterogeneity. Included by themselves in a regression specification results in a varying intercept term, which is analysis of variance (ANOVA) as a regression problem. Multiplying a covariate X by them [i.e., creating interaction variables with Hadamard (Styan, 1973) products[4] of vectors] for inclusion in a regression specification results in a varying regression slope coefficient. This specification becomes ANCOVA as a regression problem. The covariate X can be factored from all interaction variates in which it appears to determine how its regression slope coefficient varies across subgroups. These well-established and time-honored techniques furnish a fixed effects basis for conceptualizing SVCs.

[4] In matrix algebra, the Hadamard product is a cell-by-cell procedure that operates on two matrices of the same dimensions, producing a third matrix with these initial dimensions by multiplying the pairs of (i, j) elements in the first two matrices to calculate the (i, j) element entry in the third matrix.

For MESF, rather than multiply random variable X by an indicator variable, X is multiplied by eigenvectors (Griffith, 2008). Relating this procedure to specifications in the preceding section, the Hadamard product of the vector **1** and a set of eigenvectors yields the candidate set for constructing a spatially varying intercept term. Inclusion of these interaction variables dramatically increases the size of the ESF candidate selection set. Accordingly, stepwise selection can proceed only in its forward[5] (not strictly backward) mode when the total number of candidate and original variables approaches, equals, or exceeds n.

To illustrate this geographic varying regression coefficients specification, consider the relationship between median household income (MHI) and population density (PD); one expectation is that urban areas have higher PDs and higher MHIs than rural areas. The Box–Cox transformations to improve the normality of these two measures for Texas counties in 2010 are the logarithm for PD and.

$$1 - [(\text{MHI} + 35,521)/10,000]^{-1.21}$$

for MHI (subtracting the power-transformed term from one preserves the positive relationship between PD and MHI; the initial correlation is 0.27, whereas the correlation between the transformed variables is 0.25); these transformed variates may be denoted by, respectively, TRPD and TRMHI. Table 6.2 reports a summary of conventional and MESF SVC estimation results. Accounting for SA impacts on the variance estimates, which is well known. The data contain considerable redundant geographic information (about 75%). SA biases the parameter estimates because of the presence of common map pattern components (a multicollinearity effect).

The stepwise regression procedure, employing a 0.10 level of significance, results in the selection of 19 PSA and 14 NSA eigenvectors (from a candidate set of $65+89=154$) for the intercept coefficient, and 10 PSA and 11 NSA eigenvectors for the slope coefficient. The n-by-1 vectors of varying coefficients $\mathbf{a}_v$ and $\mathbf{b}_v$ are of the following form, where the groups identified by parentheses designate PSA and NSA components:

[5] Although this situation does not preclude each forward step being followed by a backward step for only those variates included in the regression equation at a given step.

Table 6.2 Summary results for a constant and a SVCs bivariate linear regression analysis: Y=TRPD, and X=TRMHI

Coefficient type	a	Standard error of a	b	Standard error of b	R^2	Shapiro–Wilk probability
Constant	0.91250	0.00188	0.00167	0.00039	0.0661	0.0355
Variable	0.91469	0.00116	0.00110	0.00025	0.8123	0.2037

Note: The spatially varying a and b values are the means across the Texas geographic landscape; the SVCs deviate by county from these global values.

$$\begin{aligned}\mathbf{a}_v = [&\mathrm{a}\mathbf{1} + (\mathrm{a}_1\mathbf{E}_1 + \mathrm{a}_6\mathbf{E}_6 + \mathrm{a}_7\mathbf{E}_7 + \mathrm{a}_{10}\mathbf{E}_{10} + \mathrm{a}_{12}\mathbf{E}_{12} + \mathrm{a}_{13}\mathbf{E}_{13} + \mathrm{a}_{14}\mathbf{E}_{14} \\ &+ \mathrm{a}_{16}\mathbf{E}_{16} + \mathrm{a}_{18}\mathbf{E}_{18} + \mathrm{a}_{20}\mathbf{E}_{20} + \mathrm{a}_{22}\mathbf{E}_{22} + \mathrm{a}_{23}\mathbf{E}_{23} + \mathrm{a}_{24}\mathbf{E}_{24} + \mathrm{a}_{25}\mathbf{E}_{25} \\ &+ \mathrm{a}_{30}\mathbf{E}_{30} + \mathrm{a}_{39}\mathbf{E}_{39} + \mathrm{a}_{55}\mathbf{E}_{55} + \mathrm{a}_{58}\mathbf{E}_{58} + \mathrm{a}_{60}\mathbf{E}_{60}) + (\mathrm{a}_{167}\mathbf{E}_{167} \\ &+ \mathrm{a}_{171}\mathbf{E}_{171} + \mathrm{a}_{175}\mathbf{E}_{175} + \mathrm{a}_{177}\mathbf{E}_{177} + \mathrm{a}_{187}\mathbf{E}_{187} + \mathrm{a}_{189}\mathbf{E}_{189} + \mathrm{a}_{207}\mathbf{E}_{207} \\ &+ \mathrm{a}_{212}\mathbf{E}_{212} + \mathrm{a}_{217}\mathbf{E}_{217} + \mathrm{a}_{224}\mathbf{E}_{224} + \mathrm{a}_{234}\mathbf{E}_{234} + \mathrm{a}_{244}\mathbf{E}_{244} \\ &+ \mathrm{a}_{245}\mathbf{E}_{245} + \mathrm{a}_{249}\mathbf{E}_{249})] \circ \mathbf{1}, \end{aligned} \tag{6.2}$$

and

$$\begin{aligned}\mathrm{b}_v = [&\mathrm{b}\mathbf{1} + (\mathrm{a}_5\mathbf{E}_5 + \mathrm{a}_9\mathbf{E}_9 + \mathrm{a}_{11}\mathbf{E}_{11} + \mathrm{a}_{18}\mathbf{E}_{18} + \mathrm{a}_{28}\mathbf{E}_{28} + \mathrm{a}_{36}\mathbf{E}_{36} \\ &+ \mathrm{a}_{44}\mathbf{E}_{44} + \mathrm{a}_{52}\mathbf{E}_{52} + \mathrm{a}_{58}\mathbf{E}_{58} + \mathrm{a}_{62}\mathbf{E}_{62}) + (\mathrm{a}_{167}\mathbf{E}_{167} + \mathrm{a}_{170}\mathbf{E}_{170} + \mathrm{a}_{198}\mathbf{E}_{198} \\ &+ \mathrm{a}_{199}\mathbf{E}_{199} + \mathrm{a}_{200}\mathbf{E}_{200} + \mathrm{a}_{206}\mathbf{E}_{206} + \mathrm{a}_{212}\mathbf{E}_{212} + \mathrm{a}_{219}\mathbf{E}_{219} + \mathrm{a}_{223}\mathbf{E}_{223} \\ &+ \mathrm{a}_{229}\mathbf{E}_{229} + \mathrm{a}_{249}\mathbf{E}_{249})] \circ \mathbf{X}, \end{aligned} \tag{6.3}$$

where $\circ$ denotes Hadamard matrix multiplication (i.e., element-by-element multiplication). Fig. 6.2 portrays these two SVCs (Fig. 6.2A and D) as well as the PSA and NSA components (respectively, Fig. 6.2B and E, and Fig. 6.2C and F). One noticeable difference between MESF and GWR results is that MESF SVCs tend to exhibit more fragmented map patterns, and GWR SVCs tend to exhibit much smoother map patterns.

Based on the confidence interval for the global coefficient, these SVCs can be classified as being significantly less than, not significantly different from, and significantly greater than their global constant value counterparts. Fig. 6.3 portrays results of this classification. The spatially varying intercept range is given by the interval (0.87135, 0.96415); Fig. 6.3A indicates that most counties are either significantly less than (*dark green*) or greater than (*dark red*) the global value of 0.91469. Meanwhile, the spatially varying slope range is given by the interval (−0.00885, 0.00838). Fig. 6.3B indicates that a number of counties have a statistically stronger relationship (*dark red*)

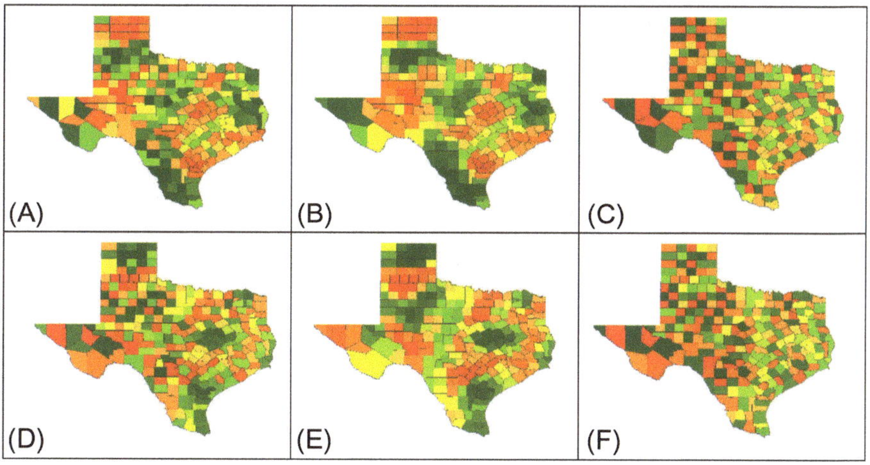

Fig. 6.2 MESF SVCs for Texas PD as a function of median household income; values are directly proportional to the scale *dark green* to *yellow* to *dark red.* (A) The regression intercept (MC = 0.44; GR = 0.56). (B) The PSA intercept component (MC = 0.74; GR = 0.27). (C) The NSA intercept component (MC = −0.43; GR = 1.37). (D) The regression slope (MC = 0.13; GR = 0.88). (E) The PSA slope component (MC = 0.72; GR = 0.32). (F) The NSA slope component (MC = −0.46; GR = 1.45).

between TRPD and TRMHI than is represented by the global slope estimate. Furthermore, counties where the relationship is significantly less than (*dark green*) the constant slope value can have either no or even a negative relationship between these two variates. The state's major metropolitan regions are conspicuous in both maps.

This simple bivariate SVCs regression equation performs reasonably well. At 0.81, its R^2 value is relatively high. In terms of cross-validation, the ratio of its predicted residual error and regression error sums of squares (PRESS/ESS) statistic is 1.62, which is quite respectable, especially given its accompanying high R^2 value. Its residuals have a Shapiro–Wilk normality diagnostic statistic probability of 0.20, and a heteroscedasticity test statistic (White, 1980) probability of 0.40 (i.e., fail to reject the null hypothesis that the residuals have homogeneous variance). The final equation has 34 terms in its spatially varying intercept and 22 terms in its spatially varying slope; the residual SA MC z-score is 1.44 (i.e., the residuals contain only a trace amount of SA). Finally, its coefficient variance inflation factors (VIFs) do not exceed 1.60, which suggests that multicollinearity is not a serious problem.[6]

[6] Hadamard products of eigenvalues and covariates are not necessarily mutually orthogonal and uncorrelated.

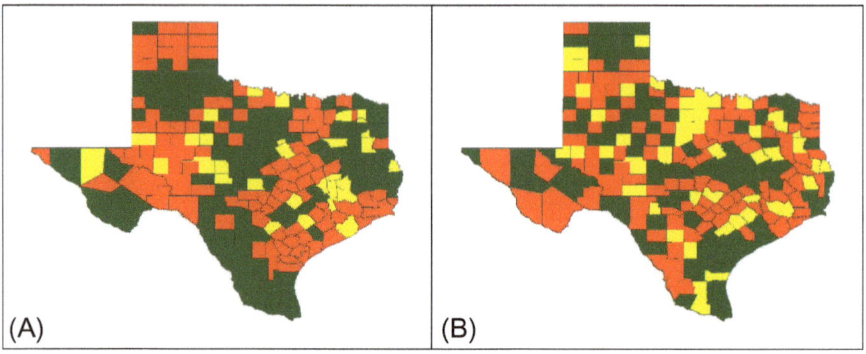

Fig. 6.3 Intercept and slope value changes across Texas: *dark green* denotes values significantly less than their global value, *yellow* denotes values not significantly different from their global value, and *dark red* denotes values significantly greater than their global value. (A) The spatially varying intercept. (B) The spatially varying slope.

6.3 Multicollinearity in spatially varying coefficients

Multicollinearity is a complicating feature of SVCs. One of it sources is the presence of common eigenvectors in these coefficients; such common eigenvectors inflate, whereas unique eigenvectors suppress, a correlation. The modified spatial weights matrix (SWM) eigenvectors are mutually orthogonal and uncorrelated (Section 2.1.2), preventing any within coefficient correlation. The two preceding SVCs [Eqs. (6.1) and (6.2)] have only the following four eigenvectors in common: $\mathbf{E}_{18}$, $\mathbf{E}_{167}$, $\mathbf{E}_{212}$, and $\mathbf{E}_{249}$. Here the outcome of these common vectors is a correlation between the spatially varying intercept and slope of -0.70 (Fig. 6.4A), which is strong but not excessive. Fig. 6.4B and C demonstrate that no empirical relationship exists between either the spatially varying intercept or slope variates and the covariate MHI; this is an attractive feature for this MESF specification.

6.4 Local SA ESFs

The preceding sections emphasize heterogeneity, a violation of the IID assumption of classical statistics. That preceding chapter (i.e., Chapter 5) addresses this issue in terms of SVCs specifications, an ANCOVA type of conceptualization, that allow local relationships to vary across a geographic landscape rather than posit single landscape-wide global relationships. This conception is in the tradition of the statistical identification of outliers, leverage observations, and influential observations, as well as nonparametric statistical

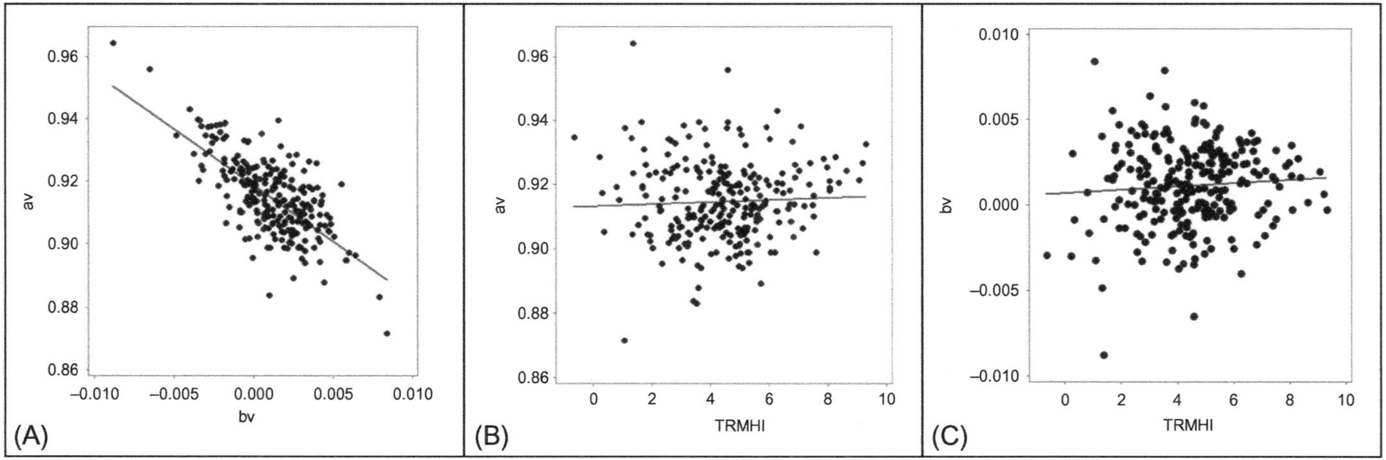

Fig. 6.4 Collinearity scatterplots for 2010 Texas TRPD and TRMHI. (A) The SVCs. (B) The spatially varying intercept versus the single covariate. (C) The spatially varying slope versus the single covariate.

calculations and/or smoothing using localized subsets of data [e.g., locally weighted scatterplot smoothing (LOESS, or LOWESS), kernel estimation; e.g., Cleveland, 1979; Simonoff, 1996]. Griffith (2003, pp. 125–128) discusses relationships between these types of diagnostic statistics and MESF.

SA is a critical feature of georeferenced data and, not surprisingly then, has been subjected to decompositions of its global indices into local indices (Anselin, 1995; Getis & Ord, 1992, 2001; Lloyd, 2007; Ord & Getis, 1995; Sokal, Oden, & Thompson, 1998). One formulation devised by Anselin is the local Moran coefficient (MC), which relates to the Moran scatterplot, much like the traditional regression notions of leverage and influential observations.[7] An analogous formulation is the local Geary ratio (GR), which alludes to a geostatistical perspective (see Section 1.2.2) that furnishes the foundation for the Getis–Ord G_i statistics. One also could imagine local join count statistics for categorical georeferenced data.

This section summarizes conceptual discussions of local SA statistics in the context of MESF, and its contents are useful in two ways: (1) ESFs constructed for local SA statistics can furnish insights into the georeferenced data multiple testing problem, and (2) relationships between ESFs and local SA statistics can be informative about the existence of mixtures of PSA and NSA.

6.4.1 Local versus global SA

Similar to a Pearson product moment correlation coefficient, the MC and the GR provide averages of n individual SA measures for a collection of n areal units forming a map. The IID assumption implies a Moran scatterplot whose cloud of points ideally forms a concentric dispersion about a bivariate axes origin, with roughly equal numbers of points located in each of the four quadrants of the plane. One feature of a well-behaved Moran scatterplot (i.e., one whose graphical portrayal makes the SA it depicts obvious by inspection) is that its scatter of points (z_i, $\sum_{j=1}^{n} c_{ij} z_j$) for a given SWM **C** forms a cloud that concentrates around its trend line, with at least most of the points within, and perhaps a few just beyond the boundaries of, its prediction interval. Fig. 6.5 exemplifies these contentions. Fig. 6.5A portrays the Moran scatterplot for a random permutation of the 2010 log-population density by county in Texas (see Fig. 3.1A). The widest bands are the

[7] In regression analysis, leverage refers to the degree to which an X value is an outlier (i.e., remote) in terms of the scatterplot cloud of points and hence distorts summary statistics and possibly the slope of the accompanying trend line; neighboring X values are missing. Meanwhile, an *influential* point is one that disproportionately influences any outcome of a regression analysis, including distorting the slope of the accompanying trend line.

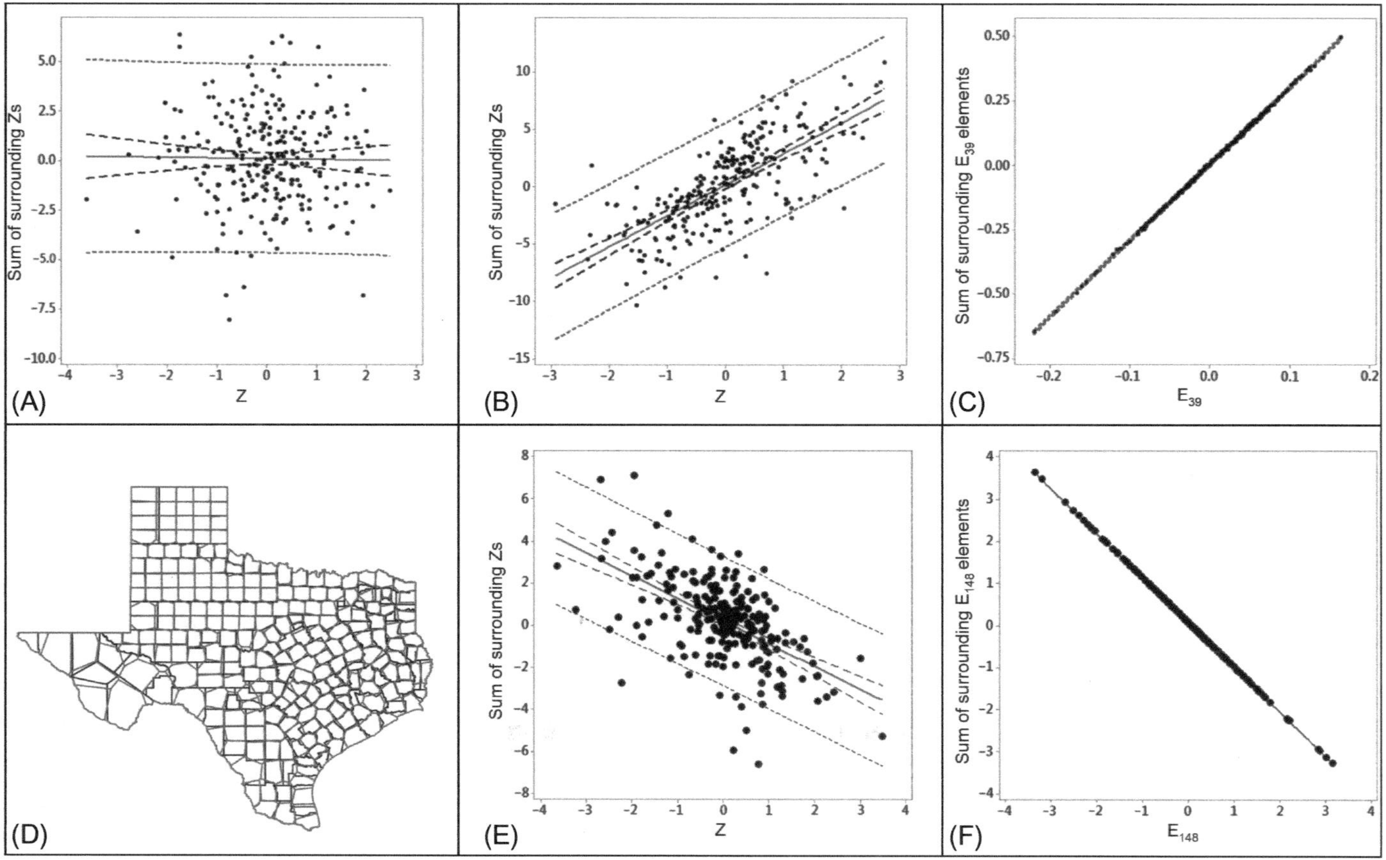

Fig. 6.5 Selected Moran scatterplots with confidence (*long dashed* lines) and prediction (*short dashed* lines) intervals. (A) Geographically random data. (B) 2010 Texas PD. (C) $\mathbf{E}_{39}$. (D) Counties with superimposed county centroid-based Thiessen polygons. (E) County-to-Thiessen polygon area ratio. (F) $\mathbf{E}_{148}$.

prediction intervals, which contain most of its data points; those data points outside of these intervals are relatively close to its boundaries. The zero slope trend line indicates zero SA. Fig. 6.5B portrays the Moran scatterplot for the 2010 log-population density (MC = 0.523). Again, the widest bands, the prediction intervals, contain almost all of its data points. In this case, those data points outside of these prediction intervals tend to be closer to its boundaries than most deviant points in Fig. 6.5A. Fig. 6.5C portrays the Moran scatterplot for eigenvector E_{39}, whose SA is moderate and positive (MC = 0.528), selected here because its MC is similar that for the 2010 log-population density; it is approximately the same as the SA underlying Fig. 6.5B. In this third case, the scatter of points aligns with, rather than deviates from, the trend line, and, consequently, the confidence and prediction intervals coincide with this trend line. This third scatterplot highlights that the MC value relates not only to deviations from a trend line but also to the slope of that trend line. These same features pertain to the case of NSA. The ratio of county-to-Thiessen polygon area in Fig. 6.5D exemplifies NSA (MC = −0.206; GR = 1.162). Fig. 6.5F portrays the Moran scatterplot for eigenvector E_{148}, whose NSA is comparable (MC = −0.204).

Another feature of a well-behaved Moran scatterplot is that, for PSA, few points outside of its prediction interval are located in the second and fourth quadrants [low-high (LH) and high-low (HL), respectively; see Section 1.2.2]; in Fig. 6.5B, only six are located in these parts of the graph. For NSA, few points would be located outside of the prediction interval in the first and third quadrants [high-high (HH) and low-low (LL), respectively]; in Fig. 6.5E, only 13 are located in these parts of the graph.

One noteworthy complication here concerns the range of SA measures for a given SWM. For the Texas SWM based on rook adjacencies, the range of NSA is (−0.635, −0.004), whereas the range for PSA is (−0.004, 1.098); the rescaled extreme eigenvalues—(254)(5.7147)/1322=−0.6346 and (254)(−3.3028)/1322=1.0980—of the modified SWM $(\mathbf{I}-\mathbf{11}^T/n)\mathbf{C}(\mathbf{I}-\mathbf{11}^T/n)$ determine these endpoints (see Section 2.1.4), whereas the intermediate point is given by $-1/(n-1)$. In other words, the NSA range (i.e., 0.631) is much narrower than the PSA range (i.e., 1.102); it needs to be stretched to reach a lower limit of −1, whereas the PSA range needs to be shrunk to reach an upper limit of 1, so that both ranges are the same, namely one. In relative terms, the NSA observed here is somewhat comparable in magnitude to that of the 2010 Texas PD PSA [i.e., $(-0.206+0.004)/(-0.635+0.004)\approx 0.320$; $(0.523+0.004)/(1.098+0.004)\approx 0.483$]. Griffith (2017, pp. 170–171) proposes a solution to this complication that needs additional research attention.

Spatial analysts employ local SA statistics to identify clusters in geographic space. Of particular interest are clusters of HH values (i.e., points located in the first quadrant of the Moran scatterplot), which often are labeled hot spots. Clusters of LL values (i.e., points located in the third quadrant of the Moran scatterplot), which often are labeled cold spots, also are of interest. Clusters of LH and HL (i.e., points located in the second and fourth quadrants of the Moran scatterplot) represent NSA, if they deviate noticeably from the origin and constitute one of the few empirical situations in which NSA regularly materializes.

6.4.2 Local MCs for ESFs

Local MCs, namely MC_is, are based on Eq. (1.1). Each part of the MC numerator summation becomes an individual statistic denoted by MC_i:

$$z_i \sum_{j=1}^{n} c_{ij} z_j, i = 1, 2, \ldots, n;$$

in other words, the horizontal and vertical axis terms in a Moran scatterplot (see Section 1.2.2) are multiplied (i.e., the cross-product nature of the MC). The sum of a set of these MC_i values needs to be divided by the quantity $\mathbf{1}^T\mathbf{C1}$ to obtain its corresponding global MC value. Because one of these calculations is for each areal unit, a map can be constructed with a set of MC_is, and then statistically significant MC_is can be identified on this map. To construct the n z-score test statistics.

$$z_{MC_i} = \frac{MC_i - E(MC_i)}{\sqrt{VAR(MC_i)}},$$

for SWM $\mathbf{C}$, under the null hypothesis of zero SA, the expected value of MC_i, $E(MC_i)$, is

$$-\frac{\sum_{j=1}^{n} c_{ij}}{n-1} = -\frac{n_i}{n-1},$$

(Anselin, 1995, p. 99), where n_i denotes the number of geographic neighbors for areal unit i, and the variance of MC_i, $VAR(MC_i)$, assuming normality,[8] is

[8] The bivariate linear regression for the ArcMap z-scores as the response variable and these approximate z-scores (z_A) as the covariate, calculated for the 2010 Texas log-population density variable, is $-0.000 + 1.006z_A$, $R^2 \approx 1$.

$$\frac{n_i(n-3)}{n-1} + 2\frac{6-n}{(n-1)(n-2)}$$

(after Anselin, 1995, p. 99). This statistical significance needs to account for multiple testing to avoid false cluster discoveries. A relatively simple way to achieve this goal is to employ a Bonferroni adjustment (e.g., Bland & Altman, 1995): divide the significance level by some quantity, which traditionally is either n or n_i (the default in R module *spdep*) for local spatial statistics (e.g., LISA). The consequence is a change in the critical value, which moves further away from zero. This particular adjustment tends to shrink cluster sizes, causing some clusters to vanish. Table 6.3 reports the way this critical value changes with adjustment between $n=1$ and $n=10{,}000$, which would constitute a large number of administrative or Thiessen polygon areal units, and $n=1{,}000{,}000{,}000$ (a billion), which would constitute a very large remotely sensed image size. The z-scores appearing in this table are not so large that a researcher would not anticipate encountering them, especially for images such as Fig. 2.1. This Bonferroni adjustment tends to be rather conservative when division is by n (i.e., the overall Type I error rate is smaller than α), and rather liberal when division is by one (i.e., the overall Type I error rate is α). Two suggested alternatives are to divide by n_i (an option in R) or by n^* (the effective sample size; Griffith, 2005; Acosta, Vellajos, & Griffith, 2018). Based on simulation experiments (see Appendix 6.A; Figs. 6.A.1 and 6.A.2, and Table 6.A.1 summarize the simulation experiment results), the preferred adjustment is based on n^*.

MESF offers two procedures for understanding MC_is. The first accounts for SA in their geographic distribution by constructing an ESF. The second separates aspatial information from map pattern, allowing map pattern information alone to furnish the basis for geographic clusters (i.e., the calculation of MC_is). Fig. 6.6A portrays the geographic distribution of statistically significant (at the 5% level) MC_i z-scores. The HH clusters (*light red*) identify the major metropolitan regions of Dallas, Austin–San Antonio, and

Table 6.3 The absolute value of z-score critical values for commonly used α

Level of significance (α)	n = 1	n = 10,000	n = 1,000,000,000
0.10	1.28155	4.26489	6.36134
0.05	1.64485	4.41717	6.46695
0.025	1.95996	4.56479	6.57094
0.01	2.32635	4.75342	6.70602
0.005	2.57583	4.89164	6.80650

Note: Probabilities rather than z-scores are used in this chapter for classification purposes.

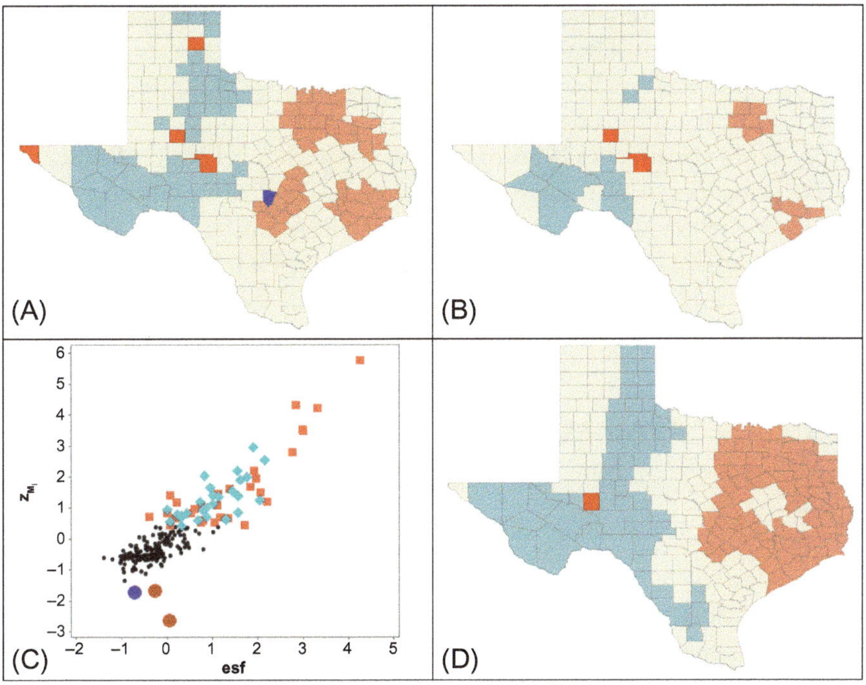

Fig. 6.6 MC_i analysis results for the 2010 Texas log-population density; *light red* denotes HH, light blue denotes LL, dark blue denotes LH, and *dark red* denotes HL. (A) 5% level of significance LISA. (B) Bonferroni-adjusted LISA (0.272% level of significance). (C) A scatterplot of MC_is versus the constructed ESF. (D) A 5% level of significance LISA ESF.

Houston. The LL clusters (*light blue*) identify more sparsely populated regions of the state. As anticipated, only a few (i.e., three) small clusters relate to NSA. The LH (*dark blue*) cluster highlights Kenedy County, between Brownsville and Corpus Christi, whereas HL (*dark red*) clusters highlight El Paso and San Angelo. Fig. 6.6B portrays geographic cluster shrinkage attributable to employing a Bonferroni adjustment based on n* (i.e., 0.05/36.75).[9] In this case, the MC_is have moderate PSA (MC = 0.57; GR = 0.48), implying that the standard Bonferroni adjustment (i.e., division by n) is too severe (see Chun and Griffith, 2013, pp. 93–94). The constructed ESF contains 36 PSA and two NSA eigenvectors, and accounts for roughly 78% of the geographic variance in the MC_is. Fig. 6.6C portrays a scatterplot of the MC_i z-scores versus the ESF. The significant

[9] Section 3.2 reports the SAR SA parameter estimate, $\hat{\rho} = 0.70120$. Substituting this value together with $n = 254$ into the equation in Griffith (2005, 743) yields $n^* \approx 36.75$.

HH and LL PSA clusters are the extreme data points extending into the first quadrant of the graph. The significant LH and HL NSA clusters are the extreme negative z-scores (these are the only ones located outside of the 95% prediction interval accompanying a bivariate regression analysis).

Fig. 6.6D portrays the LISA map (based on a 5% level of significance) for the 2010 log-population density ESF (Fig. 3.1B). The smoothing process eliminates the NSA and most single-county PSA clusters, and expands or modifies the larger PSA clusters. The individual HH and LL clusters are preserved; the three major metropolitan regions of the state remain visually conspicuous.

6.4.3 Local GRs for ESFs

Anselin (1995, p. 100) also posits the local GR, namely GR_i, as $\sum_{j=1}^{n} c_{ij}\left(y_i - y_j\right)^2$, an expression that, in this form, does not support the construction of a SA graphic analogous to the Moran scatterplot (see Section 1.2.2). Earlier, Griffith (1978, p. 44) formalized the algebraic relationship between the global MC and GR, which later was rewritten in a simplified form as follows:

$$GR = \frac{n-1}{\sum_{i=1}^{n}\sum_{j=1}^{n} c_{ij}} \frac{\sum_{i=1}^{n}\left[\left(\sum_{j=1}^{n} c_{ij}\right)(y_i - \bar{y})^2\right]}{\sum_{i=1}^{n}(y_i - \bar{y})^2} - \frac{n-1}{n} MC \qquad (6.4)$$

(Chun and Griffith 2013, p. 12).[10] This equation implies that a Geary scatterplot can be constructed by measuring z_i on the horizontal axis, as is done for a Moran scatterplot, and measuring the quantity

$$z_i \sum_{j=1}^{n} c_{ij} - \sum_{j=1}^{n} c_{ij} z_j$$

on the vertical axis (Fig. 6.7A). Accordingly, a version of GR_i may be defined in terms of the quantity

$$z_i \left[z_i \sum_{j=1}^{n} c_{ij} - \sum_{j=1}^{n} c_{ij} z_j \right]. \qquad (6.5)$$

[10] The equation appearing in this source has a typographical error; an erroneous 2 appears in the denominator of the first term on the right-hand side of the equation.

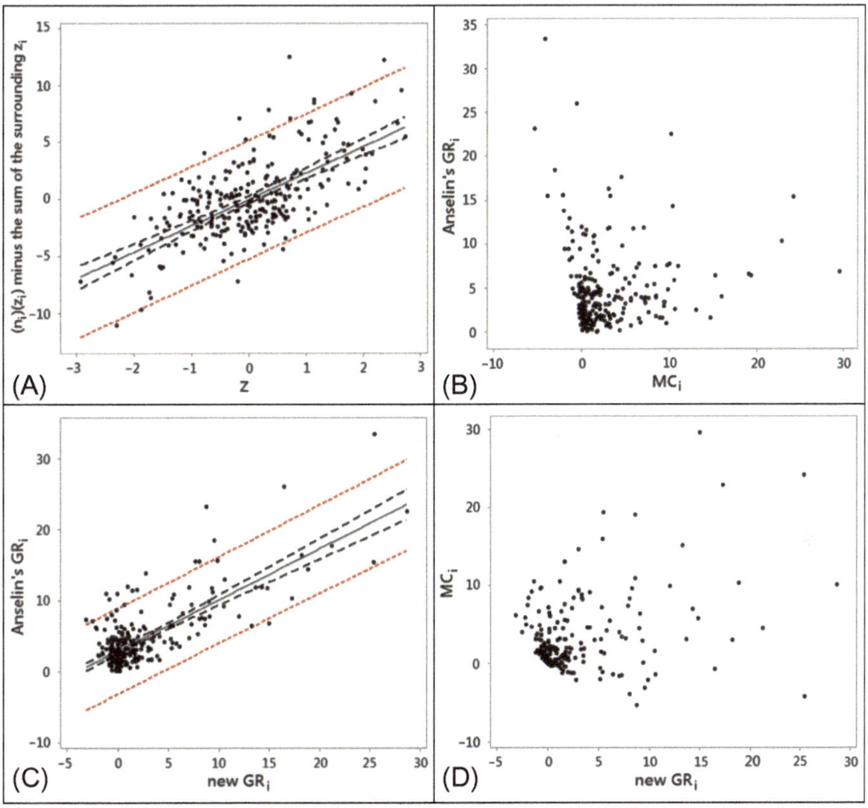

Fig. 6.7 Local GR scatterplots for the 2010 log-population density (MC = 0.52; GR = 0.44). (A) A new Geary scatterplot; the *solid gray line* denotes the trend line, the *long dashed lines* denote the 95% confidence intervals, and the *short dashed lines* denote the 95% prediction intervals. (B) The relationship between the MC_i and Anselin's GR_i. (C) The new GR_i versus Anselin's GR_i. (D) The new GR_i versus the MC_i.

Similar to a set of MC_i values, the sum of a set of these GR_i values needs to be divided by the quantity $\mathbf{1}^T\mathbf{C}\mathbf{1}$ to obtain its corresponding global GR value. One drawback of this local index is that it can yield a negative value; both the local and global GRs are nonnegative when expressed in terms of squared real numbers. Another weakness is that the expected value of Eq. (6.4) is n_i, which can be a relative large integer.

Both Figs. 1.6B and 6.7A exhibit linear relationships. Neither the plot of Anselin's GR_i index nor the square root of his GR_i index versus z-scores renders a useful scatterplot; this feature highlights a strength of Eq. (6.4). The relationship between MC_i and Anselin's GR_i (Fig. 6.7B) implies that these two indices may furnish different information about local SA.

Fig. 6.7C suggests that a conspicuous, but very weak, linear relationship exists between Anselin's GR_i and Eq. (6.4). Fig. 6.7D implies that Eq. (6.4) also may furnish different information about local SA.

Sokal et al. (1998) formulate another version of GR_i, namely

$$\frac{\sum_{j=1}^{n} c_{ij}\left(y_i - y_j\right)^2}{\sum_{i=1}^{n}(y_i - \bar{y})^2/n}.$$

Converting the attribute variables in a dataset to z-scores translates this expression into

$$\frac{n}{n-1}\sum_{j=1}^{n} c_{ij}\left(z_i - z_j\right)^2,$$

which is a rescaled version of Anselin's GR_i index. Its sampling distribution constructed with randomization has an expected value of

$$\frac{2nn_i}{n-1},$$

and, assuming a normally distributed random variable, has a variance of

$$\frac{6nn_i(n_i-1)}{n-1} - \left(\frac{2nn_i}{n-1}\right)^2.$$

These two terms support the calculation of z-score test statistics equivalent to those for MC_i.

An ESF description of Anselin's GR_i index[11] constructed for the 2010 Texas log-population density contains 21 PSA and one NSA eigenvectors, and accounts for roughly 46% of the geographic variance in these local indices across the state. An ESF description of Eq. (6.4) contains seven PSA and 14 NSA eigenvectors, and accounts for roughly 34% of the geographic variance. GR_i maps are not presented here because no statistically significant cluster exists. An important weakness of the GR is that its variance is greater than that for the MC, again implying that the latter is preferable to the form as a SA index.

The selected GR_i weaknesses highlighted in this section furnish insights into why local GR statistics are not popular. The algebraic Eq. (6.4) relating the global

[11] A GR_i index value was calculated with 2010 log-population density data for each of the 254 Texas counties. These values constitute a response variable, and stepwise linear regression using a candidate set of eigenvectors as the covariates enables an ESF to be constructed (see Chapter 3) for local GR statistics.

MC and GR indices implies that MC captures much of the SA information in a dataset, with GR merely emphasizing impacts of outlier attribute values or locations distorting a SWM with a disproportionate number of neighbors. Furthermore, MESF furnishes only partial descriptions of the SA in these local statistics, because its eigenvectors come from a modified SWM **C**, whereas the spatial structure affiliated with a GR comes from a different SWM specification, one that relates to a Laplacian matrix (Bollobás, 2013), which in turn relates to the standard **W** SWM. Research to date implies that Geary eigenvector spatial filtering (GESF) possesses poor properties vis-à-vis MESF and is not preferable to it (Griffith, 2017; Griffith & Li, 2017). Consequently, evidence presented in this section supports the contention that local GR values furnish spatial analysts with little value added beyond that obtainable with local MC values.

6.4.4 Local Getis–Ord statistics for ESFs

The Getis–Ord G_i statistic (Getis & Ord, 1992, 2001; Ord & Getis, 1995) is another local SA index that has captured spatial analysts' attention. It also has a global counterpart but is restricted to nonnegative attribute variables; its advantage over the MC_i is its ability to differentiate between hot and cold spots. This index has two versions: G_i, which does not include the attribute value at location i, and G_i^*, which does include the attribute value at location i. The equation for the former version is

$$G_i = \frac{\sum_{j=1}^{n} c_{ij} y_j}{\sum_{j=1}^{n} y_j}, i \neq j$$

(i.e., location i not included in either its numerator or denominator), where $E(G_i) = \frac{\sum_{j=1}^{n} c_{ij}}{n-1}$; its variance formula is rather complex.

Ord and Getis (1995, p. 289), and Chun and Griffith (2013, p. 97), among others, present the following z-score version of this index:

$$\frac{\sum_{j=1}^{n} c_{ij} y_j - \bar{y} \sum_{j=1}^{n} c_{ij}}{s_y \sqrt{\frac{(n-1) \sum_{j=1}^{n} c_{ij} - \left(\sum_{j=1}^{n} c_{ij} \right)^2}{n-2}}}.$$

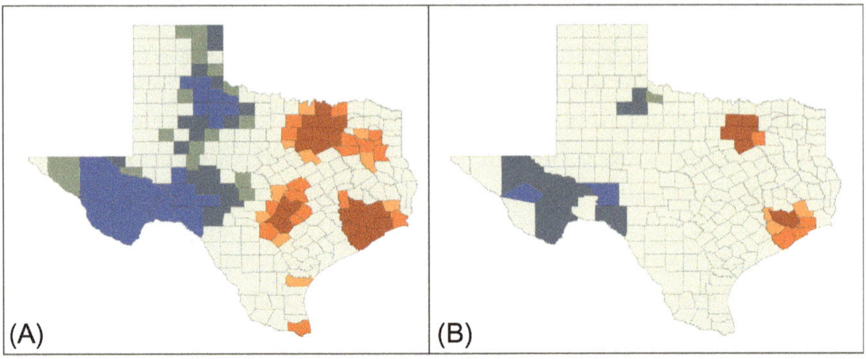

Fig. 6.8 Local Getis-Ord G_i statistics for log-population density. *Blue* denotes cold spots, and *red* denotes hot spots: dark denotes 99%, medium denotes 95%, and light denotes 90% confidence intervals. (A) Unadjusted. (B) A Bonferroni adjustment based on n*.

Positive G_i values indicate clusters of relatively high attribute values, whereas negative G_i values indicate clusters of relatively low attribute values. This index also benefits from Bonferroni-adjusted significance tests. Fig. 6.8 presents results for the 2010 log-population density attribute variable. As before, the three major metropolitan regions of Texas are conspicuous, here as hot spots. Shrinkage from a Bonferroni adjustment fails to eliminate Dallas and Houston as hot spots. Cold spots are in west Texas, with two remaining (Jeff Davis and Crockett Counties) after a Bonferroni adjustment.

The relationship between the MC_i and the G_i is quadratic in nature (Fig. 6.9A) and is strong (the linear regression with a first- and second-order term renders $R^2=0.83$). This nonlinear relationship signifies why the G_i

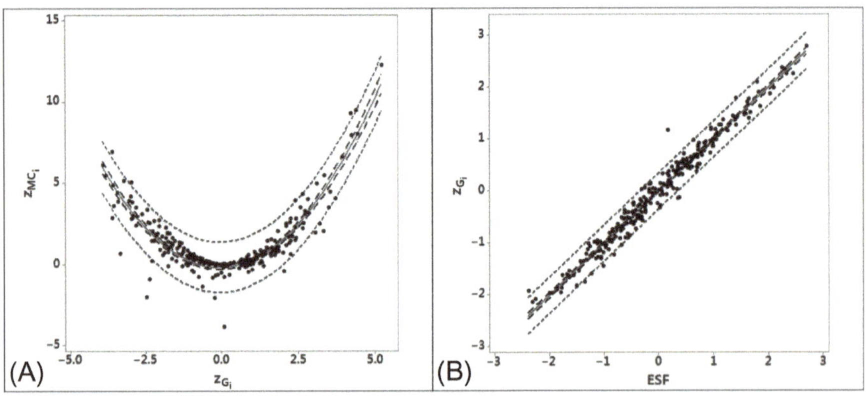

Fig. 6.9 Visualizations of G_i: the *gray solid curve* denotes the trend line, the *black long dashed curves* denote the 95% confidence interval, and the *short dashed curves* denote the 95% prediction interval. (A) The relationship between the MC_i and the G_i. (B) The ESF versus the full set of G_i values.

index successfully differentiates between hot and cold spots, whereas the MC_i successfully uncovers outliers. These Texas log-population density-based G_i indices display strong positive SA (MC = 0.91; GR = 0.13). Their ESF contains 48 PSA and four NSA eigenvectors, and accounts for roughly 97% of their geographic variance. Fig. 6.9B portrays a scatterplot of the G_i z-scores versus the ESF: this scatterplot has a much tighter scatter of points about its trend line than that for the corresponding MC_is (compare Figs. 6.9B and 6.5C). Only three of the G_i values are outside of the prediction interval (Fig. 6.9B). In other words, the geographic distribution of these G_is exhibits considerable structure.

6.5 Summary

This chapter presents an alternative to SVCs produced by GWR-inspired or RE model specifications. These alternative coefficients are centered on their global constant coefficient counterparts (Moran eigenvectors have a mean of zero). Any multicollinearity these SVCs contain can be traced to their common eigenvectors. Their geographic visualizations tend to be fragmented rather than artificially smooth. They more closely relate to the RE-based SVCs because they align with the SSRE component of a RE. This chapter highlights how to adapt ANCOVA and interaction regression variables to extend MESF. Once constructed, interaction terms for a slope are selected by a stepwise forward selection procedure; the common covariate simply needs to be factored from them, with the resulting linear combination of eigenvectors, an ESF, furnishing the spatially varying slope. The mechanics of this type of analysis reinforces the contention that a simple ESF constitutes a spatially varying intercept term. One of the major advantages of MESF here is that it exploits the SA statistical distribution theory for linear regression residuals developed by Cliff and Ord (1973, 1981). Finally, the set of SVCs produced by MESF supports prediction; the PD-MHI example presented in this chapter achieves an R^2 that exceeds 0.8, when its conventional aspatial bivariate regression value is less than 0.1.

This chapter also presents discussion about relationships between local indices of SA and latent SA in the map patterns of these indices, with special reference to local MC_is and Getis–Ord G_is, characterizing this latent SA with ESFs. This overall treatment suggests several themes for fruitful future research, including using ESFs to help devise appropriate multiple comparison methodology, an unresolved problem in this part of spatial statistics. This chapter also presents evidence that local GR_is may not be very informative, at least not beyond local MC_is. This theme also needs considerable subsequent research. In addition, local SA analyses need to be extended to join count statistics, as advocated for in this chapter, and perhaps even to spatial autoregressive ρ parameters. Nevertheless, this chapter demonstrates that MESF furnishes insights about local measures of SA, especially the Getis–Ord G_i index. As such, MESF should become a standard part of local SA analyses, and an important component of spatial heterogeneity.

Appendix 6.A Bonferroni adjustment simulation experiment results[12]

Ignoring the multiple testing problem is equivalent to dividing the Type I error probability, α, by one; one outcome is a sizable increase in the risk of false positives (identifying clusters that do not exist), essentially with probability one. The traditional Bonferroni adjustment yields statistically conservative results; one outcome is an increase in the risk of false negatives (not identifying clusters that exist). Because the number of neighbors, n_i, tends not to be greater than about 35 in empirical examples (Griffith, 2018), based on a rook definition of adjacency, the default R Bonferroni adjustment deviates only modestly from no adjustment. The effective sample size[13] tends to be both substantially greater than even the maximum n_i and substantially less than n (see Griffith 2005, 2015, p. 2169). It furnishes a trade-off between conservative and liberal statistical results (Figs. 6.A.1 and 6.A.2 and Table 6.A.1).

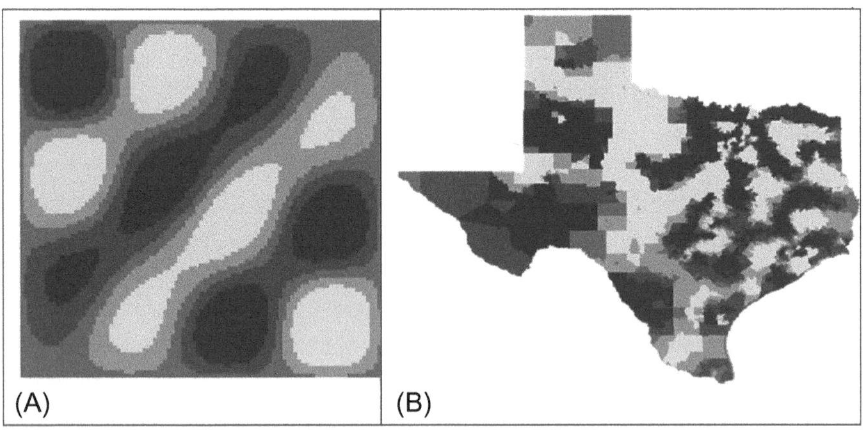

Fig. 6.A.1 Two specimen geographic distributions with clusters. (A) A 100-by-100 regular square tessellation, MC≈0.994. (B) The 158,112,010 census block groups of Texas, MC≈0.956.

[12] The authors are indebted to Ms. Lan Hu, a doctoral student at the University of Texas at Dallas during the writing of this book, for conducting the simulation experiments.

[13] Effective sample size, n*, here is based on the spatial statistical SAR model specification.

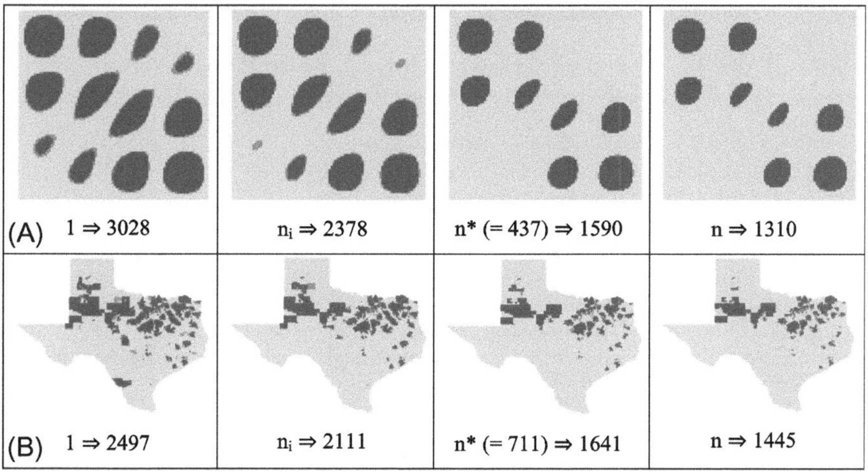

Fig. 6.A.2 LISA clusters identified for the specimen surfaces portrayed in Fig. 6.A.1, $\alpha=0.05$. (A) Fig. 6.A.1A. (B) Fig. 6.A.1B.

Table 6.A.1 The average number of statistically significant LISA clusters (rounded to the nearest integer), rook definition of adjacency, $\alpha=0.05$, 1000 replications

		No adjustment	Bonferroni adjustment		
Regular square lattice	ρ_{SAR}	**(i.e., 1)**	n_i	n*	n
20-by-20	0.10	19	10	3	3
	0.50	35	24	12	10
	0.90	85	66	61	38
	0.95	97	77	72	46
100-by-100	0.10	483	258	35	34
	0.50	889	584	183	154
	0.90	2120	1643	1064	735
	0.95	2417	1915	1288	920
500-by-500	0.10	12,056	6419	459	444
	0.50	22,163	14,503	2945	2533
	0.90	52,933	41,006	19,619	14,296
	0.95	60,452	47,865	24,461	18,369

References

Acosta, H., Vellajos, R., & Griffith, D. (2018). On the effective geographic sample size. *Journal of Statistical Computation and Simulation*, *88*, 1958–1975.

Anselin, L. (1995). Local indicators of spatial association—LISA. *Geographical Analysis*, *27*, 93–115.

Besag, J. (1974). Spatial interaction and the statistical analysis of lattice systems. *Journal of the Royal Statistical Society B*, *36*, 192–236.

Bland, J., & Altman, D. (1995). Multiple significance tests: The Bonferroni method. *The BMJ*, *310*, 170.
Bollobás, B. (2013). *Modern graph theory* (corrected ed.). Berlin: Springer-Verlad.
Chun, Y., & Griffith, D. (2013). *Spatial statistics and Geostatistics*. Thousand Oaks, CA: SAGE.
Cleveland, W. (1979). Robust locally weighted regression and smoothing scatterplots. *Journal of the American Statistical Association*, *74*, 829–836.
Cliff, A., & Ord, J. (1973). *Spatial Autocorrelation*. London: Pion.
Cliff, A., & Ord, J. (1981). *Spatial processes*. London: Pion.
Getis, A., & Ord, J. (1992). The analysis of spatial association by use of distance statistics. *Geographical Analysis*, *24*, 189–206.
Getis, A., & Ord, J. (2001). Testing for local spatial autocorrelation in the presence of global spatial autocorrelation. *Journal of Regional Science*, *41*, 411–432.
Griffith, D. (1978). *Spatial autocorrelation: A primer*. Washington, DC: AAG.
Griffith, D. (2003). *Spatial autocorrelation and spatial filtering: Gaining understanding through theory and scientific visualization*. Berlin: Springer-Verlag.
Griffith, D. (2005). Effective geographic sample size in the presence of spatial autocorrelation. *Annals of the AAG*, *95*, 740–760.
Griffith, D. (2008). Spatial filtering-based contributions to a critique of geographically weighted regression (GWR). *Environment and Planning A*, *40*, 2751–2769.
Griffith, D. (2013). Estimating missing data values for georeferenced Poisson counts. *Geographical Analysis*, *45*, 259–284.
Griffith, D. (2015). Approximation of Gaussian spatial autoregressive models for massive regular square tessellation data. *International Journal of Geographical Information Science*, *29*, 2143–2173.
Griffith, D. (2017). Some robustness assessments of Moran eigenvector spatial filtering. *Spatial Statistics*, *22*, 155–179.
Griffith, D. (2018). Generating random connected planar graphs. *GeoInformatica*, 22, 767–782.
Griffith, D., & Li, B. (2017). A geocomputation and geovisualization comparison of Moran and Geary eigenvector spatial filtering. In: *CPGIS Publication Committee, Proceedings of the 25th international conference on geoinformatics, Geoinformatics 2017*, SUNY/Buffalo, Buffalo, NY, August 2–4.
Lloyd, C. (2007). *Local models for spatial analysis*. Boca Raton, FL: CRC Press.
Ord, J., & Getis, A. (1995). Local spatial autocorrelation statistics: Distributional issues and an application. *Geographical Analysis*, *27*, 286–306.
Patuelli, R., Griffith, D., Tiefelsdorf, M., & Nijkamp, P. (2011). Spatial filtering and eigenvector stability: Space–time models for German unemployment data. *International Regional Science Review*, *34*, 253–280.
Simonoff, J. (1996). *Smoothing methods in statistics*. Berlin: Springer.
Sokal, R., Oden, N., & Thompson, B. (1998). Local spatial autocorrelation in a biological model. *Geographical Analysis*, *30*, 331–354.
Styan, G. (1973). Hadamard products and multivariate statistical analysis. *Linear Algebra and its Applications*, *6*, 217–240.
Tiefelsdorf, M., & Griffith, D. (2007). Semi-parametric filtering of spatial autocorrelation: The eigenvector approach. *Environment and Planning A*, *39*, 1193–1221.
Waller, L., Zhu, L., Gotway, C., Gorman, D., & Gruenewald, P. (2007). Quantifying geographic variations in associations between alcohol distribution and violence: A comparison of geographically weighted regression and spatially varying coefficient models. *Stochastic Environmental Research and Risk Assessment*, *21*(5), 573–588.
White, H. (1980). A heteroskedasticity-consistent covariance matrix estimator and a direct test for heteroskedasticity. *Econometrica*, *48*, 817–838.

CHAPTER 7

Spatial interaction modeling

Chapter outline

Chapters 1–6 deal with static georeferenced data; Chapter 8 deals with space–time data, moving the discussion toward dynamic georeferenced data. In both cases, each datum is tied to a single location (i.e., point or areal unit polygon). This chapter addresses flows in geographic space (i.e., spatial interaction), tying each datum to a pair of locations, one labeled an origin and the other a destination. Static georeferenced data contain n observations (i.e., the number of locations). Space–time data contain nT observations, for T points in time. Spatial interaction data contain n(n-1) observations for movement across areal unit boundaries (i.e., within areal unit movement is ignored) and n^2 observations for combined inter- and intra-areal unit movement.

Describing spatial interaction with an equation has a long and storied history, with Carrothers (1956), Sen and Smith (1995), and Farmer and Oshan (2017), among many others, furnishing point-in-time reviews of this topic in terms of its popular gravity model formulation, whose general form is as follows:

$$F_{ij} = g\left[O_i, D_j, f\left(d_{ij}\right)\right], \tag{7.1}$$

where F_{ij} denotes the flow between origin i and destination j areal units, O_i is some attribute variable(s) quantifying relevant features of origin i, D_j is some attribute variable(s) quantifying relevant features of destination j, g is some function (frequently multiplicative in form), and $f(d_{ij})$ is some deterrent

https://doi.org/10.1016/B978-0-12-815043-6.00007-0

function of the distance[1] separating origin i and destination j (frequently inverse power or negative exponential in form). The initial functional form resulted in Eq. (7.1) being transformed to a log-linear version and a normally distributed error attached to it for estimation and inference purposes, yielding

$$\mathrm{LN}\left(F_{ij}+\delta\right)=\mathrm{LN}(\kappa)+\alpha\mathrm{LN}(O_i)+\beta\mathrm{LN}\left(D_j\right)-\gamma d_{ij}+\varepsilon_{ij},$$

where LN denotes natural logarithm, ε_{ij} is a normally distributed random error term, $\delta \geq 0$ added to the response variable F_{ij} is a translation parameter needed if any $F_{ij}=0$, and κ, α, β, and γ are linear regression coefficients; sometimes this equation includes $\mathrm{LN}(d_{ij})$ rather than d_{ij}. This specification results in exponents for the O_i and D_j variables when the equation is back-transformed to its multiplicative form. Difficulties associated with this implementation include that F_{ij} often is a count for which many cases are zero (hence the need for δ). Furthermore, the back-transformation of the logarithm form results in the error structure being multiplicative, rather than additive, in nature: $e^{\varepsilon_{ij}}$. For the cases of flows that are counts, Flowerdew and Aitkin (1982) show that the correct implementation is to retain Eq. (7.1) in a multiplicative form, and to treat F_{ij} as a Poisson random variable (RV). Meanwhile, Wilson (1967) shows that, in many situations, the deterrence function should be a negative exponential one [favoring $\mathrm{LN}(d_{ij})$, rather than simply d_{ij}, in the preceding equation]. He also establishes that, rather than exponents for the O_i and D_j variables (hence, $\alpha=\beta=1$), multiplicative weights A_i and B_j should be included in the definition of function g. These weights are balancing factors ensuring that the total of the predicted flows equals the total of the observed flows both leaving origin i and entering destination j.

Working with a logit model of joint origin–destination choice, Anas (1983) indirectly links these balancing factors to a Poisson specification with fixed effects such that each origin and each destination has its own 0–1 indicator variable (i.e., 2n binary variables, each having a single one and n-1 zeros); the individual log-balancing factor values are proportional to the indicator variable regression coefficients. Griffith and Fischer (2013) furnish

[1] For simplicity, Euclidean distance is employed here that is calculated with latitude and longitude coordinates. This distance definition is reasonable because Texas is not excessively large vis-à-vis the entire Earth and hence may be viewed as a flat surface. A bivariate linear regression of great-circle distance on this Euclidean distance for the 32,385 Texas distance pairs yields an intercept of 0.03127, a slope of 1.0010, and an R^2 of 0.9999. In other words, these simpler distances introduce little measurement error into the analyses summarized in this chapter.

proofs of these equivalencies in general, and also that they are equivalent to random effects (see Section 6.4) having spatially structured (SSRE) and spatially unstructured (SURE) components. Curry (1972) argues that specifications like Eq. (7.1) fail to account for latent spatial autocorrelation (SA) in georeferenced flows data. LeSage and Pace (2008) effectively address this SA issue, but for the log-linear version of Eq. (7.1) using spatial Bayesian techniques. Griffith (2009) effectively addresses this issue for the Poisson specification of Eq. (7.1) containing balancing factors by employing Moran eigenvector spatial filtering (MESF) methodology. Griffith, Fischer, and LeSage (2017) compare these implementations. Griffith (2011) emphasizes how MESF supports the visualization of SA in spatial interaction data.

Griffith (2009) describes and analyzes the 2000 Texas intercounty journey-to-work data, which furnishes some background for the data analyses presented in this chapter. Meanwhile, this chapter describes and analyzes the 2010–14 Texas intercounty migration data, as well as the 2009–13 intercounty journey-to-work data. Here $n^2 = 254^2 = 64{,}516$, with 84.7% of the migration, and 90.0% of the journey-to-work, origin–destination pairs having a flow of zero.

7.1 Initial spatial interaction descriptions of internal Texas migration

During the five-year period 2010–14 in Texas, 1,031,966 people moved across county borders, and 2,627,883 moved within county borders. The simplest description of these flows is the classic gravity model version of Eq. (7.1),[2] namely

$$F_{ij} = \kappa\, O_i^{\alpha} D_j^{\beta} e^{-\gamma d_{ij}} + \varepsilon_{ij}, \tag{7.2a}$$

where α and β are exponential parameters, respectively, indicating the importance of the origin and destination quantities, γ is the distance decay parameter, κ is a constant of proportionality, and ε_{ij} is an i-j dyad-specific random error term assumed to be a normal RV for this specification. Here, the measures of O_i and D_j are the sums of out- and in-migration for the 2010–14 time period under study. To avoid zero-count difficulties (e.g.,

[2] Original restrictive computer capabilities and the availability of only normal curve theory for estimation purposes resulted in a multiplicative error and parameter estimation by employing linear regression and converting Eq. (7.2a) to a log-linear form. Eq. (7.2a) can retain normal curve theory for estimation purposes but yields better results with a nonlinear regression specification that includes an additive error term.

taking the logarithm of zero), Eq. (7.2a) parameters were estimated with nonlinear regression. Table 7.1 reports parameter estimates for this specification.[3]

The Poisson RV description of flows yields fitted values calculated as follows:

$$E(F_{ij}) = n_T p_{ij},$$

where E is the calculus of expectations operator, n_T is the total number of flows in a system (e.g., Texas), and p_{ij} is the probability of migrating from origin areal unit i (e.g., a county) to destination areal unit j. For a Poisson RV, p_{ij} is defined as a Poisson probability, and one result is that the sums of the fitted and observed flows are the same (see Table 7.1). This equation also implies the following modified version of Eq. (7.2a):

$$F_{ij} = n_T \frac{O_i^{\alpha} D_j^{\beta} e^{-\gamma d_{ij}}}{\sum_{j=1}^{n} \sum_{i=1}^{n} O_i^{\alpha} D_j^{\beta} e^{-\gamma d_{ij}}} + \varepsilon_{ij}. \quad (7.2b)$$

The fractional part of this equation defines migration probability p_{ij}.

The description of these flows involving balancing factors is the following version of Eq. (7.1):

$$F_{ij} = \kappa A_i O_i B_j D_j e^{-\gamma d_{ij}} + \varepsilon_{ij}, \quad (7.3)$$

where A_i and B_j are the balancing factors operating as surrogates of unspecified socioeconomic/demographic covariates of migration; a Poisson regression requires $LN(O_i) + LN(D_j)$ to be included in a regression analysis as an offset variable (i.e., its coefficient is set to one rather than estimated). This specification is known as the doubly constrained spatial interaction model. Table 7.1 reports parameter estimates for its specification with the Texas migration data.

A comparison of the parameter estimates for Eqs. (7.2b)-Poisson and (7.3) indicates that corresponding ones are somewhat—although not drastically—different. Assuming a Gaussian (i.e., normal) rather than a Poisson RV introduces specification error resulting in noticeable bias, both in parameter estimates (e.g., the estimated exponents differ by a factor of

[3] Letting the attributes equal the origin and destination populations renders $\hat{\kappa} = e^{-4.0378}$, $\hat{\alpha} = 0.5956$, $\hat{\beta} = 0.5351$, $\hat{\gamma} = 0.0824$, and $R^2 = 0.9859$. These estimation results are very similar to those for equation (7.2a) based upon flow totals.

Table 7.1 Simple spatial interaction model parameter estimates: Texas migration, 2010–14

Equation	$\hat{\kappa}$	$\hat{\alpha}$	$\hat{\beta}$	$\overline{A}_i$	$\overline{B}_j$	$\hat{\gamma}$	R^2	Over-dispersion	$\frac{\sum_{i=1}^{n}\sum_{j=1}^{n}\hat{F}_{ij}}{\sum_{i=1}^{n}\sum_{j=1}^{n}F_{ij}}$
(7.2a)-normal	$e^{-1.3290}$	0.5763	0.5103	1	1	0.0817	0.9956		0.899
(7.2b)-Poisson	$e^{-5.7759}$	0.7092	0.7015	1	1	0.0302	0.9290	9.0126	1.000
(7.3)	$e^{-9.7872}$	1	1	0.9376	0.8080	0.0287	0.9429	8.3620	1.000
(7.5)	$e^{-12.8481}$	1	1	0.9744	1.0181	0.0121	0.9953	5.0329	1.000

Note: To simplify estimation, if $O_i=0$ (King County) or $D_j=0$ (Loving County)—only a single case of each is present—then these totals were set to one, which is within the American Community Survey (ACS) margin of error (MOE) for their other totals. The next smallest totals, respectively, are 23 and 10, and the resulting respective A_i and B_j estimates essentially are zero.

two, and the distance decay exponents differ by a factor of three) and predicted values. The Poisson RV has considerable overdispersion that cannot be accounted for by switching to a negative binomial RV (see Section 5.4.2), in part because of the excessive number of zeros. The R^2 values close to one mostly are attributable to the presence of a few very large migration flows. Except for its R^2 value, the unconstrained gravity model tends to furnish a poorer description of migration flows.

Table 7.1 suggests that model quality assessment is difficult here. A small number of flows tend to force accompanying R^2 values to be relatively close to one. The estimation of probabilities used to distribute the total number of flows results in the ratio of a sum of fitted and observed flows being equal to one. This criterion dismisses only the traditional unconstrained gravity model—Eq. (7.2a)-normal—as a preferred specification. Another criterion to which Eq. (7.2a) alludes is the total in- and out-migration flows for each areal unit, which involves 508 comparisons. Fig. 7.1 reveals that pairwise scatterplots of model-based in- and out-migration county totals dismisses both Eqs. (7.2a)-normal and (7.2b)-Poisson, furnishing a standard that indicates Eq. (7.3) is the best equation; the respective pooled relative mean absolute errors[4] in these three cases are 1564, 3923, and 0.

7.2 Spatially autocorrelated origin and destination variables

Relevant SA may be uncovered with the geographic distributions of the A_i, O_i, B_j, and D_j attribute values. Fig. 7.2 portrays the concatenated Poisson regression coefficients for the two sets of origin and destination indicator variables, each augmented with zero for the arbitrarily deleted one (here it was for county 254, Zavala), which are equivalent to the $LN(A_i)$ and $LN(B_j)$ balancing factors. These two maps exhibit conspicuous SA [for Fig. 7.2A, the Moran coefficient (MC) = 0.769, and the Geary ratio (GR) = 0.113; and, for Fig. 7.2B, MC = 0.781, and GR = 0.120] differentiating the more urban regions in the eastern part of the state (e.g., Dallas–Fort Worth, Austin–San Antonio, and Houston) from the rest of the state.

The eigenvector spatial filter (ESF) describing $LN(A_i)$ comprises 50 eigenvectors [48 depicting positive SA (PSA), and two depicting negative SA (NSA)] that account for roughly 91% of the geographic variation (nearly

[4] The sum of the absolute differences between observed and model-based output total in- and out-migration flows divided by two.

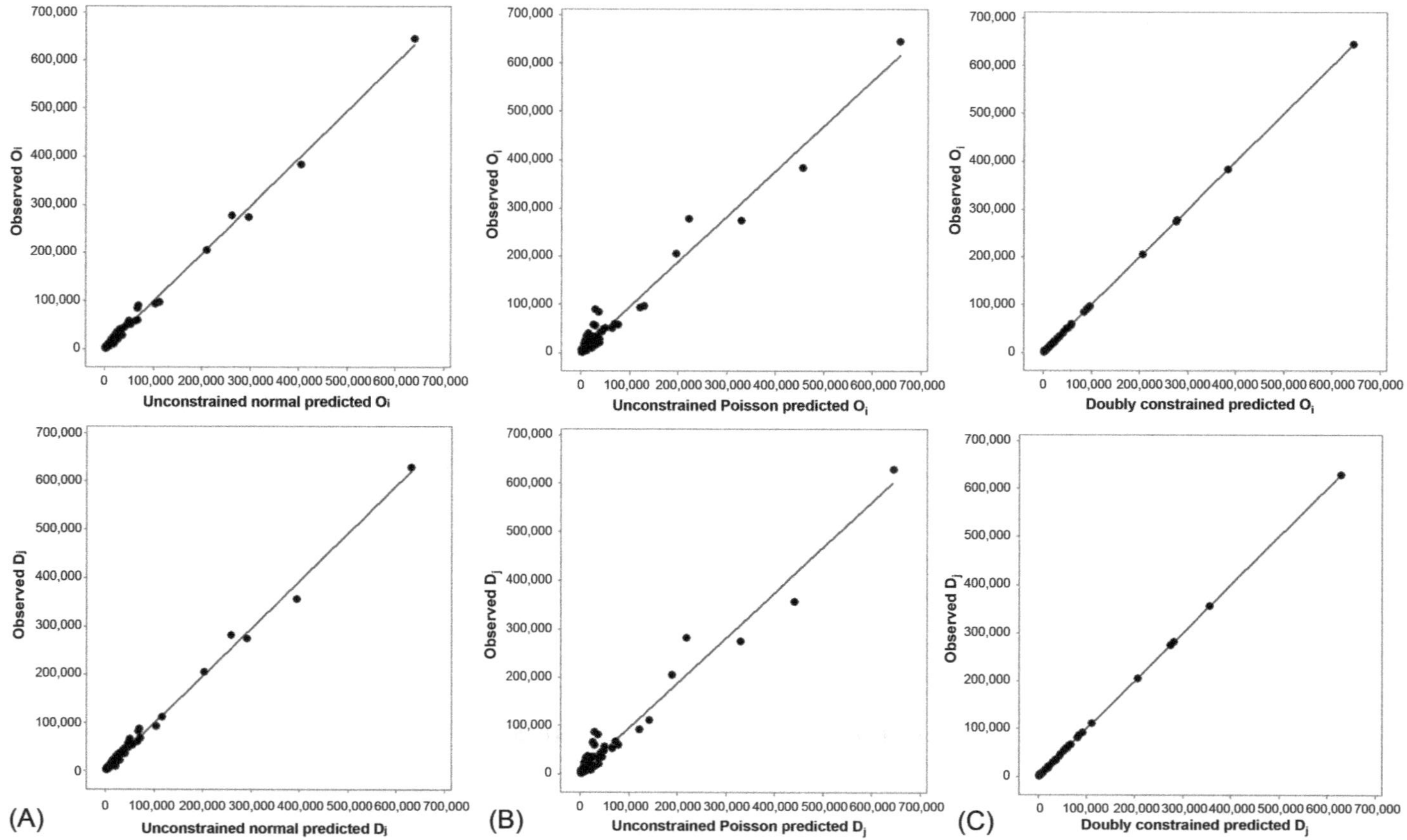

Fig. 7.1 Scatterplots portraying total origin in- (*top*) and destination out- (*bottom*) flows by county. (A) Unconstrained normal gravity model output. (B) Unconstrained Poisson gravity model output. (C) Doubly constrained gravity model output.

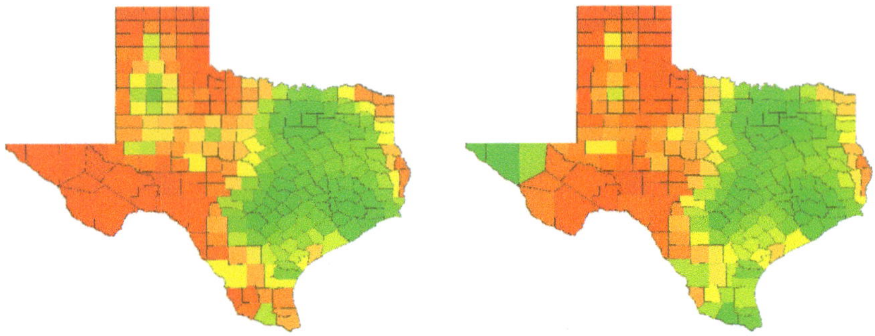

(A) (B)

Fig. 7.2 Migration balancing factor values are directly proportional to the scale *dark green* to *yellow* to *dark red*. (A) LN(A_i). (B) LN(B_j).

91% attributable to PSA, and nearly 0% to NSA) across Texas in these values. LN(A_i) has an extreme lower tail outlier but is approximately normally distributed in its Box–Cox power-transformed version $[LN(A_i) + 2.8]^{0.1}$ (with a Shapiro–Wilk normality diagnostic statistic probability of 0.02). The ESF describing LN(B_j) comprises 49 eigenvectors, all depicting PSA, that account for roughly 88% of the geographic variation in these values. LN(B_j) has deviations in both tails but conforms closely to a bell-shaped curve in its Box–Cox power-transformed version $[LN(B_j) + 3.3]^{1.2}$ (with a Shapiro–Wilk normality diagnostic statistic probability of 0.22).

Fig. 7.2 depicts the geographic distribution of migration accessibility across Texas during 2010–14 (Cesario, 1977): $1/A_i$ is an accessibility index of destinations with respect to origin i, whereas $1/B_j$ is an accessibility index of origins with respect to destination j. Respectively, A_i and B_j index relative differential migration emission and attraction qualities of origin and destination counties. Fig. 7.2 indicates that relatively high geographic accessibility mostly pertains to the Dallas–Fort Worth–Austin–San Antonio–Houston part of the state. In contrast, relatively low geographic accessibility primarily pertains to the western part of the state. Accordingly, the quantity A_iO_i indexes migration propensity of origin county i per unit of destination county accessibility; the quantity B_jD_j indexes migration propensity of destination county j per unit of origin county accessibility. Consequently, Fig. 7.2 exposes a concentration of migration origin and destination counties in the eastern part of Texas, and a dearth of migration origin and destination counties in the western part of Texas—with some additional destination attractiveness in the El Paso and the Brownsville–Corpus Christi urban regions—during the five-year time horizon under study.

7.3 Network autocorrelation in migration flows

The preceding section addresses SA in geographic attributes of either origin or destination flows, with an ESF decomposition making them explicit in a spatial interaction equation:

$$F_{ij} = \kappa e^{LN(A_i)} O_i e^{LN(B_j)} D_j e^{-\gamma d_{ij}} + \epsilon_{ij} = \kappa e^{ESF_{A_i}} e^{\xi_{A_i}} O_i e^{ESF_{B_j}} e^{\xi_{B_j}} D_j e^{-\gamma d_{ij}} + \varepsilon_{ij}, \quad (7.4)$$

(see Fischer & Griffith, 2008) where the ESF_js are equivalent to SSREs (see Section 6.1) and the ξ_js are equivalent to SUREs (see Section 8.1). However, this is not the only SA feature of spatial interaction data. An additional component may be labeled network autocorrelation (NA) and relates to dyads of areal units. Its underlying network of locations may be, for example, a regional or national urban hierarchy. Fig. 7.3 illustrates this notion for portions of the Dallas and Houston metropolitan areas. The *red arrows* denote direct flows between two locations (*black* counties), which Eqs. (7.2a)–(7.4) already capture. The adjacent *gray* counties relate to SA in geographic distributions of attributes, such as concatenated A_i and B_j values, which Eqs. (7.3) and (7.4) already capture. NA pertains to the correlations existing in geographic flows that are of three sources (Fig. 7.3B): (1) for a particular destination (denoted here by *black*), the areal units surrounding an origin (denoted here by *gray*) tend to have the size of their departure flows correlated with the magnitude of departure flows from their central location (denoted here in *black*), and hence

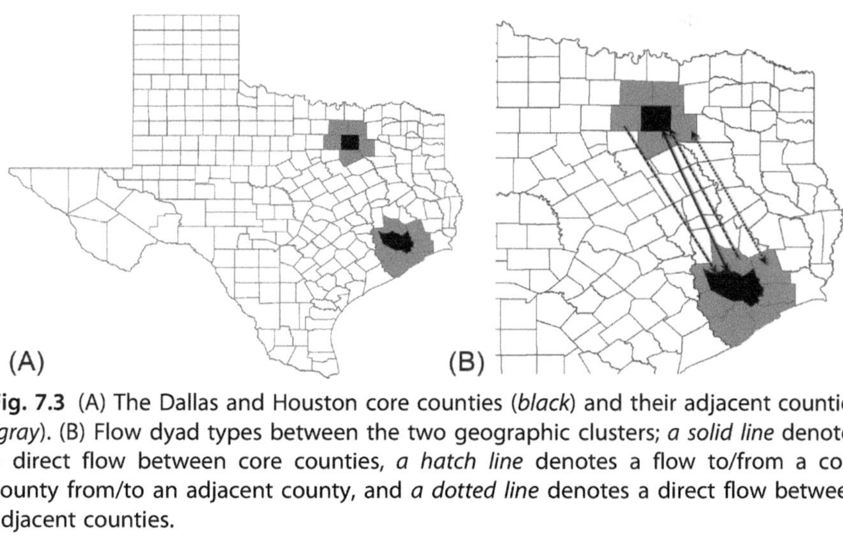

(A) (B)

Fig. 7.3 (A) The Dallas and Houston core counties (*black*) and their adjacent counties (*gray*). (B) Flow dyad types between the two geographic clusters; *a solid line* denotes a direct flow between core counties, *a hatch line* denotes a flow to/from a core county from/to an adjacent county, and *a dotted line* denotes a direct flow between adjacent counties.

each other; (2) for a particular origin (denoted here by *black*), the areal units surrounding a destination (denoted here by *gray*) tend to have the size of their arrival flows correlated with the magnitude of arrival flows to their central location (denoted here in *black*), and hence each other; and, (3) for a particular origin–destination dyad, flows between the dyads of surrounding areal unit origins and destinations tend to have the size of their flows correlated with the magnitude of the flows between their central location pair, and hence each other. None of the four equations capture this data feature.

NA introduces another term, extending the Eq. (7.4) specification to

$$F_{ij} = \kappa e^{ESF_{A_i}} e^{\xi_{A_i}} O_i e^{ESF_{B_j}} e^{\xi_{B_j}} D_j e^{-\gamma d_{ij}} e^{ESF_{ij}} + \varepsilon_{ij}, \tag{7.5}$$

where $e^{ESF_{ij}}$ describes NA impacts on migration. A Kronecker product (i.e., $\otimes$; see Section 2.4; Van Loan, 2000) of the origin and destination spatial weights matrix (SWM) n-by-1 eigenvectors furnishes the eigenvectors for this additional term (Chun & Griffith, 2011; $\mathbf{E}_i \otimes \mathbf{E}_j$ and $\mathbf{E}_j \otimes \mathbf{E}_i$); this simplification reduces the relevant eigenfunction problem size from n^2 to n.

Fig. 7.4 portrays the various SA components in the Texas migration spatial interaction data. The log-balancing factors (Fig. 7.4A and B) differ noticeably from their Eq. (7.3) counterparts (Fig. 7.2) based upon a conventional doubly constrained gravity model specification. The SA captured by these two attribute geographic distributions decreases (respectively, $MC = 0.769 \rightarrow 0.339$ and $GR = 0.113 \rightarrow 0.648$, and $MC = 0.781 \rightarrow 0.263$ and $GR = 0.120 \rightarrow 0.601$), with some SA now being captured by the NA components, which were constructed with 360 eigenvectors selected from a candidate set of 1225 eigenvectors (based upon a Kronecker product of eigenvectors from the SWM with an adjusted MC, MC/MC_{max}, of at least 0.5—thus the minimum MC for the Kronecker products is $0.5^2 = 0.25$). Nevertheless, the metropolitan regions and east–west differentiation remain conspicuous in the map patterns. The NA components (Fig. 7.4C for origins, and Fig. 7.4D for destinations) display weak to moderate PSA (for Fig. 7.4C, $MC = 0.454$ and $GR = 0.612$; and, for Fig. 7.4D, $MC = 0.465$ and $GR = 0.615$), with some metropolitan region clusters detectable in their map patterns.[5]

Fig. 7.5 presents scatterplots for observed versus fitted migration flows. Eq. (7.2a)-normal (Fig. 7.5A) displays an appealing alignment. However, its diagnostic statistics (Table 7.1) are somewhat problematic (e.g., the ratio of

[5] Because the sum of each eigenvector's elements equals zero (except for the single vector proportional to **1**), construction of the NA origin and destination ESFs uses medians.

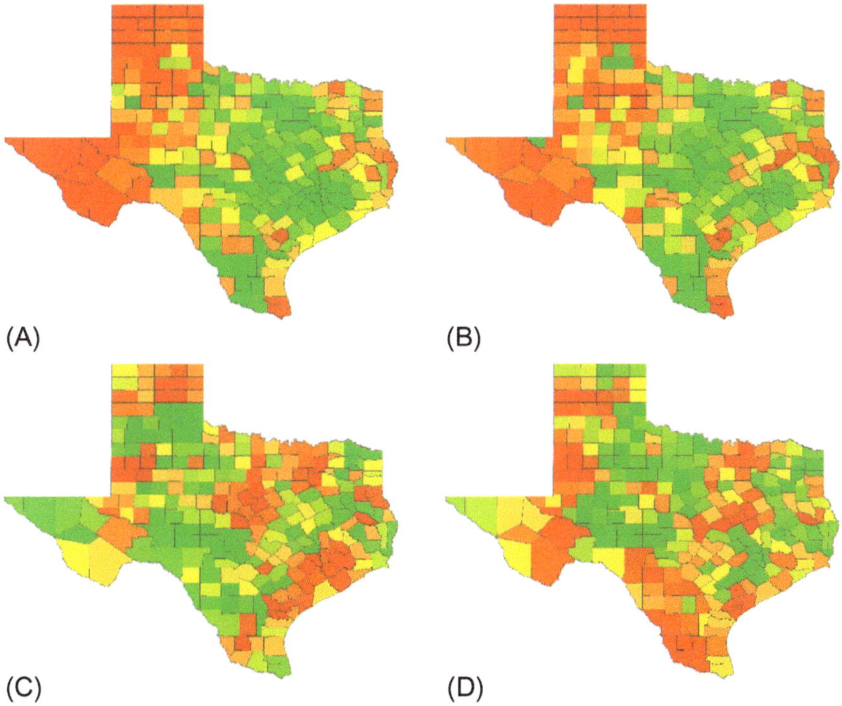

Fig. 7.4 The migration parameter values are directly proportional to a scale from *dark green* to *yellow* to *dark red*. (A) LN(A_i) adjusted for SA [see Eq. (7.5)]. (B) LN(B_j) adjusted for SA [see Eq. (7.5)]. (C) Origin NA ESF. (D) Destination NA ESF.

the total predicted and observed flows markedly deviates from one). In addition, a bivariate regression of the observed flows on its corresponding fitted flows tends to overpredict by six (the intercept). Its slope is one, as it should be (i.e., if predicted values exactly equal observed values, then a bivariate regression slope estimate for them is one). Meanwhile, although Eq. (7.3), which accounts for attribute SA, has many appealing diagnostic statistics [Table 7.1; for example, the ratio of the total predicted and observed flows is one, and the R^2 and overdispersion improve vis-à-vis Eq. (7.2a)], its scatterplot is not well aligned. Eq. (7.5), which accounts for attribute SA as well as NA, has many appealing diagnostic statistics (Table 7.1) and a well-aligned scatterplot. In addition, a bivariate regression of the observed flows on its corresponding fitted flows tends to overpredict by two (the intercept), an improvement on the Eq. (7.2a)-normal results. Its slope is 1.03, which is very close to one. Consequently, accounting for both SA and NA in spatial interaction data yields very appealing fitted flow values accompanied by more

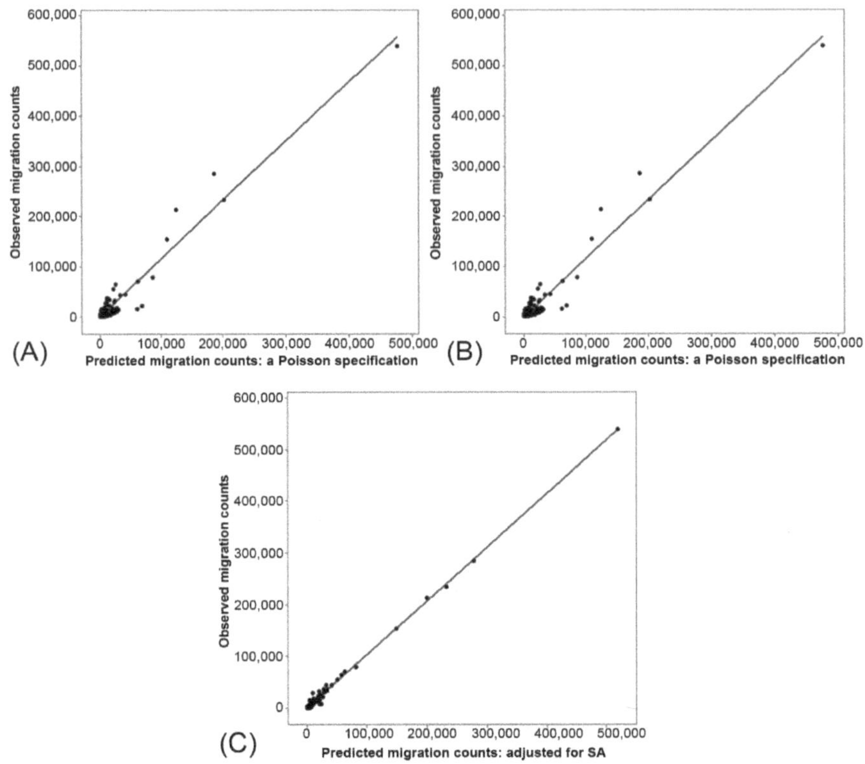

Fig. 7.5 Observed (*vertical* axis) versus fitted (*horizontal* axis) migration flow scatterplots. (A) From Eq. (7.2a)-normal. (B) From Eq. (7.3). (C) From Eq. (7.5).

ideal diagnostic statistics. The ESFs involved have a tendency to: inflate nearby flows if the NA term $e^{ESF_{ij}} > 1$, have no effect on nearby flows if the NA term $e^{ESF_{ij}} \approx 1$, and deflate nearby flows if the NA term $e^{ESF_{ij}} < 1$, with a collective impact of aligning the observed versus fitted scatterplot with a straight line both near and away from the origin of the scatterplot (see Fig. 7.5). A primary outcome is the proper estimation of the global distance decay parameter, $\hat{\gamma}$, which no longer is biased by SA and NA (i.e., increasingly accounting for SA here decreases the distance decay parameter estimate).

7.4 Spatial and network autocorrelation in journey-to-work flows: A reconnaissance

Griffith (2009) provides a parallel analysis of the 2000 Texas journey-to-work data extracted from the US decennial census, which has 9,229,209 workers traveling between their homes and their jobs, and an origin–

destination flows matrix containing 81.0% zero entries. These data are not afflicted by any zero county in- or out-flow totals. This analysis furnishes a benchmark against which comparisons may be made with results for the ACS 2009–13 journey-to-work data. These latter data have 11,319,672 workers traveling between their homes and their jobs, and an origin–destination flows matrix containing 90.0% zero entries (ACS flows are estimates, with 4021 flows smaller than their accompanying margins of error; many zero flow estimates could represent very small population numbers, resulting in a true count of zero flows closer to 81%).

Table 7.2 reports parameter estimates for the various spatial interaction gravity model specifications describing Texas journey-to-work flows; this table is analogous to Table 7.1 for migration flows. The normal approximation equation description here contains far less parameter estimation bias attributable to specification error than its migration flows counterpart (e.g., the $\hat{\alpha}$ and $\hat{\beta}$ values differ by far less), although it now has a sizable flows overprediction bias.

Fig. 7.6 portrays the concatenated Poisson regression coefficient estimates for the two sets of origin and destination indicator variables, each augmented with zero for the arbitrarily deleted one (here, as before, it is for county 254, Zavala), which are equivalent to the $LN(A_i)$ and $LN(B_j)$ balancing factors. These two maps exhibit conspicuous SA (for Fig. 7.6A, $MC = 0.643$ and $GR = 0.258$; and, for Fig. 7.6B, $MC = 0.783$ and $GR = 0.101$) differentiating the more urban region in the eastern part of the state (e.g., Dallas–Fort Worth, Austin–San Antonio, and Houston) from the rest of the state: this urban region tends to have a similar propensity, whereas the remainder of the state tends to have a dissimilar propensity, for work and jobs. SA accounts for roughly 85% of the variance in the geographic distribution of $LN(A_i)$ across Texas, whereas it accounts for about 88% for $LN(B_j)$. The ESFs for these two variates contain about three-fourths of the eigenvectors used to construct the origin and destination balancing factor ESFs for migration (see the preceding section). Again, the $LN(A_i)$ has both a PSA (a linear combination of 44 eigenvectors) and a NSA (a linear combination of six eigenvectors) component. In contrast to migration flows, $LN(B_j)$ does, too, for journey-to-work flows, although its NSA (a single eigenvector) component accounts for nearly 0% of its geographic variance; the PSA component is a linear combination of 50 eigenvectors. In other words, noticeable similarities, as well as differences, exist in the map patterns associated with migration and journey-to-work flows.

Fig. 7.7 portrays the various SA components in the Texas migration spatial interaction data. The log-balancing factors (Fig. 7.7A and B) are

Table 7.2 Simple spatial interaction model parameter estimates: Texas journey-to-work, 2009–13

Equation	$\hat{\kappa}$	$\hat{\alpha}$	$\hat{\beta}$	$\overline{A}_i$	$\overline{B}_j$	$\hat{\gamma}$	R^2	Over-dispersion	$\frac{\sum_{i=1}^{n}\sum_{j=1}^{n}\hat{F}_{ij}}{\sum_{i=1}^{n}\sum_{j=1}^{n}F_{ij}}$
(7.2a)-normal	$e^{-0.8644}$	0.4115	0.6358	1	1	0.0665	0.9873		1.124
(7.2b)-Poisson	$e^{-1.6049}$	0.4326	0.6710	1	1	0.0741	0.9849	7.5686	1.000
(7.3)	$e^{-8.8087}$	1	1	0.7069	4.0943	0.0730	0.9969	6.7344	1.000
(7.5)	$e^{-12.8481}$	1	1	0.9673	2.8162	0.0562	0.9994	4.2549	1.000

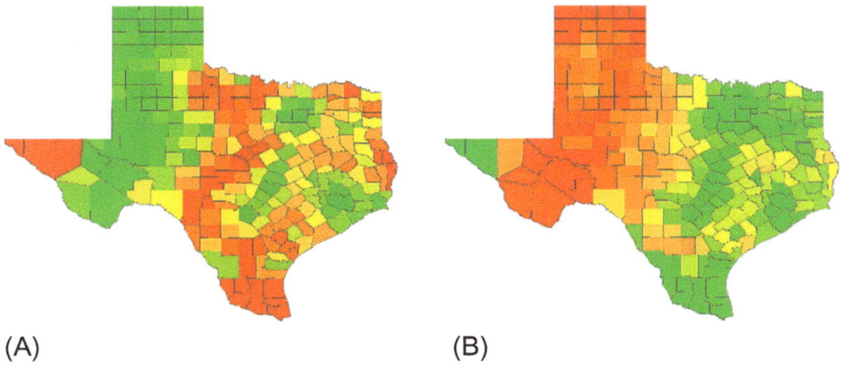

Fig. 7.6 Journey-to-work balancing factor values are directly proportional to the scale *dark green* to *yellow* to *dark red*. (A) LN(A_i). (B) LN(B_j).

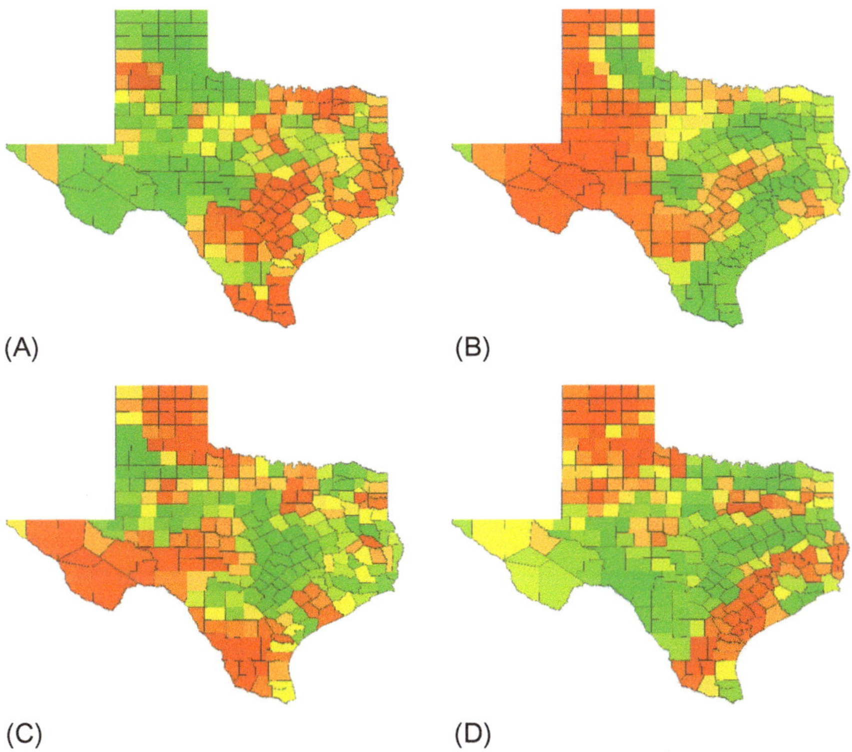

Fig. 7.7 Journey-to-work parameter values are directly proportional to a scale from *dark green* to *yellow* to *dark red*. (A) LN(A_i) adjusted for SA. (B) LN(B_j) adjusted for SA. (C) Origin NA ESF. (D) Destination NA ESF.

very similar to their Eq. (7.3) counterparts based on a conventional doubly constrained gravity model specification. The SA captured by these two attribute geographic distributions has some ambiguity, with one index indicating that it increases while the other indicates that it decreases (respectively, $MC = 0.643 \rightarrow 0.758$ and $GR = 0.258 \rightarrow 0.268$, and $MC = 0.783 \rightarrow 0.804$ and $GR = 0.101 \rightarrow 0.204$); the NA components still capture SA and were constructed with 581 eigenvectors selected from a candidate set of 1225 eigenvectors (based upon a Kronecker product of eigenvectors from the SWM with an adjusted MC of at least 0.5). Nevertheless, the metropolitan regions and east–west differentiation remain conspicuous in the map patterns. The NA components (Fig. 7.7C for origins, and Fig. 7.7D for destinations) display moderate PSA (for Fig. 7.7C, $MC = 0.675$, and $GR = 0.342$; and, for Fig. 7.7D, $MC = 0.694$, and $GR = 0.368$), with a United States–Mexico border swath for worker residence counties and a Gulf Coast swath for employment counties, detectable in their map patterns.

Paralleling Fig. 7.5, Fig. 7.8 presents scatterplots for observed versus corresponding fitted journey-to-work flows. Eq. (7.2a)-normal (Fig. 7.8A) displays an appealing alignment. However, its diagnostic statistics (Table 7.2) are somewhat problematic. Again, although Eq. (7.3), which accounts for attribute SA, has many appealing diagnostic statistics (Table 7.2), its scatterplot is not well aligned. Eq. (7.5), which accounts for attribute SA as well as NA, has many appealing diagnostic statistics (Table 7.2) and a well-aligned scatterplot. Consequently, as for migration, accounting for both SA and NA in spatial interaction journey-to-work data yields very accurate fitted flow values accompanied by more ideal diagnostic statistics. The ESFs involved have a tendency to: inflate nearby flows if the NA term $e^{ESF_{ij}} > 1$, have no effect on nearby flows if the NA term $e^{ESF_{ij}} \approx 1$, and deflate nearby flows if the NA term $e^{ESF_{ij}} < 1$, with a collective impact of aligning the observed versus fitted scatterplot with a straight line both near and away from the origin of the scatterplot (see Fig. 7.8C). As noted in the preceding section, a primary outcome of this more sophisticated model specification is the proper estimation of the global distance decay parameter, $\hat{\gamma}$, which no longer is biased by SA and NA (i.e., increasingly accounting for SA here decreases its value). Findings presented here for journey-to-work flows corroborate those summarized in Griffith (2009) for earlier Texas journey-to-work flows and those summarized in Section 7.3 for migration flows, contributing to a foundation for generalizing SA and NA results to a wide range of spatial interaction flow types.

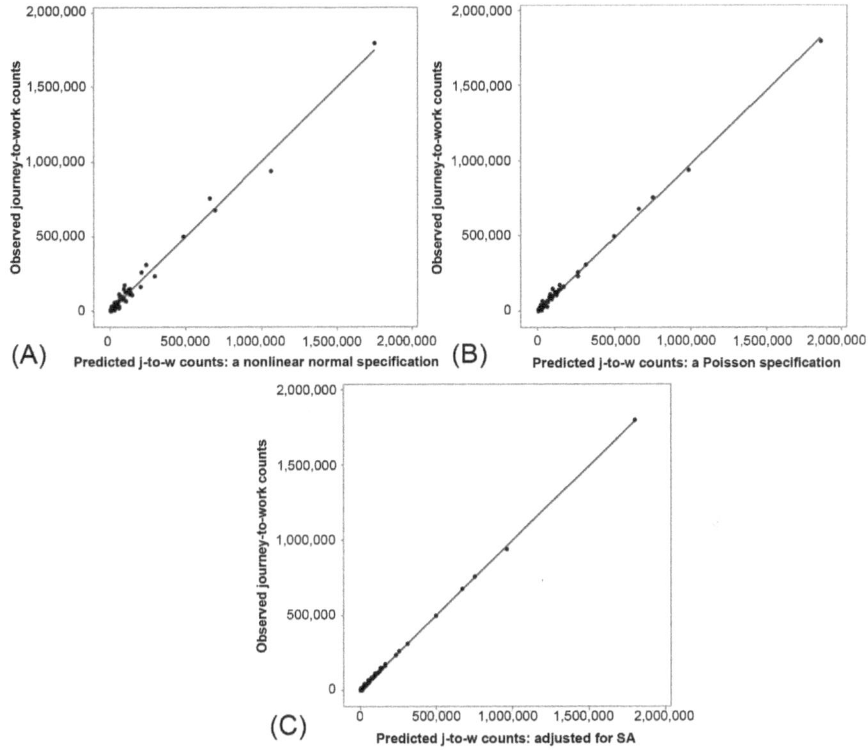

Fig. 7.8 Observed (*vertical* axis) versus fitted (*horizontal* axis) journey-to-work flow scatterplots. (A) From Eq. (7.2a)-normal. (B) From Eq. (7.3). (C) From Eq. (7.5).

7.5 A toy example: Exemplifying the necessary data structures

Because the data structure required for this type of analysis is complicated, this section presents a simple numerical example for pedagogic purposes. This example is based on 2010–14 migration flows (see Section 7.1) among the six Texas counties constituting the 2010 Corpus Christi metropolitan statistical area (MSA); $n=6$, and $n^2=36$. Fig. 7.9 presents the geographic landscape and the origin–destination migration flows matrix. Table 7.3 tabulates the data necessary to estimate the various gravity model specifications (e.g., Eqs. 7.2a-normal, 7.2b-Poisson, 7.3, and 7.5). Table 7.4 reports the SA variates. The eigenvectors are those for which a Kronecker product needs to be calculated. The balancing factor terms $LN(A_i)$ and $LN(B_j)$—which are concatenated coefficients from Fig. 7.10C augmented with zero (for the

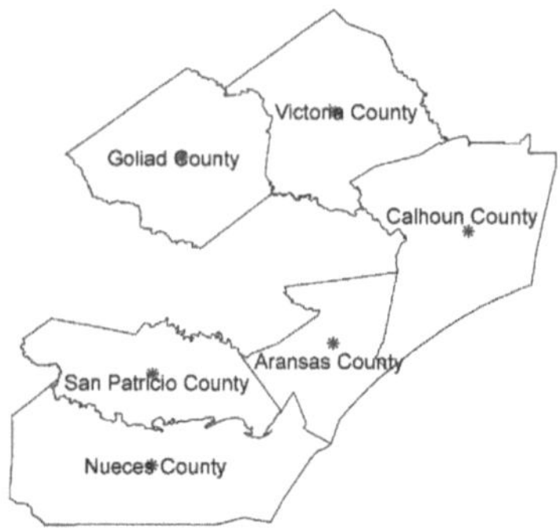

	1	2	3	4	5	6
1	5384	105	0	1544	1057	34
2	143	6696	0	26	0	939
3	0	121	1299	9	34	1164
4	168	126	0	141604	3227	292
5	651	39	9	9874	14699	73
6	21	2872	200	161	12	33774

1. Aransas County
2. Calhoun County
3. Goliad County
4. Nuece County
5. San Patricio County
6. Victoria County

(A) (B)

Fig. 7.9 (A) County centroids (asterisks). (B) Migration flows.

missing indicator variable)—are potential sources for SA in the origin and destination geographic distributions. Here eigenvectors $\mathbf{E}_1$ and $\mathbf{E}_5$ account for about 86% of the variance in LN(A_i); LN(B_j) appears to be void of SA in this example (almost certainly due to sample size). Neither of the SA-adjusted LN(A_i) and LN(B_j) correlates with any of the eigenvectors in this example (again, almost certainly due to sample size). Ten Kronecker product eigenvectors, out of 25 candidate ones (the constant eigenvector was not included), account for NA.

The following expression constitutes the NA ESF:

$$\begin{aligned}&1.4456\mathbf{E}_3 - 1.2613\mathbf{E}_5 - 2.4948\mathbf{E}_6 - 1.6473\mathbf{E}_7 + 1.8756\mathbf{E}_{11}\\ &\quad + 1.0990\mathbf{E}_{14} + 1.3703\mathbf{E}_{16} + 0.6154\mathbf{E}_{18} - 2.7289\mathbf{E}_{21} - 1.2631\mathbf{E}_{25}\end{aligned}$$

The origin and destination county medians of values calculated with this expression are the ESFs appearing in Fig. 7.11. Neither correlates with any of the eigenvectors in this example (once more, almost certainly due to sample size).

These spatial interaction data analyses cannot yet be executed with procedures in Spatial Analysis using ArcGIS Engine and R (SAAR), and ESF Tool. Appendix 7.A furnishes R code that allows replication of results for this toy example; similar code appears in Chapter 5 of Chun and Griffith (2013). The `functions.R` code appears in Appendix 7.B. This code can

Table 7.3 Dataset for estimating gravity model specifications

ID	Origin FIPS code	Destination FIPS code	Migration	Origin total (O_i)	Destination total (D_j)	Distance	Origin I_1	Origin I_2	Origin I_3	Origin I_4	Origin I_5	Destination I_1	Destination I_2	Destination I_3	Destination I_4	Destination I_5
1	4	4	5384	8124	6367	0.0000	1	0	0	0	0	1	0	0	0	0
2	4	29	105	8124	9959	33.5321	1	0	0	0	0	0	1	0	0	0
3	4	88	0	8124	1508	47.5388	1	0	0	0	0	0	0	1	0	0
4	4	178	1544	8124	153,218	40.1825	1	0	0	0	0	0	0	0	1	0
5	4	205	1057	8124	19,029	32.6467	1	0	0	0	0	0	0	0	0	1
6	4	235	34	8124	36,276	49.0638	1	0	0	0	0	0	0	0	0	0
7	29	4	143	7804	6367	33.5321	0	1	0	0	0	1	0	0	0	0
8	29	29	6696	7804	9959	0.0000	0	1	0	0	0	0	1	0	0	0
9	29	88	0	7804	1508	51.6135	0	1	0	0	0	0	0	1	0	0
10	29	178	26	7804	153,218	73.4295	0	1	0	0	0	0	0	0	1	0
11	29	205	0	7804	19,029	62.6176	0	1	0	0	0	0	0	0	0	1
12	29	235	939	7804	36,276	32.8026	0	1	0	0	0	0	0	0	0	0
13	88	4	0	2627	6367	47.5388	0	0	1	0	0	1	0	0	0	0
14	88	29	121	2627	9959	51.6135	0	0	1	0	0	0	1	0	0	0
15	88	88	1299	2627	1508	0.0000	0	0	1	0	0	0	0	1	0	0
16	88	178	9	2627	153,218	63.7990	0	0	1	0	0	0	0	0	1	0
17	88	205	34	2627	19,029	45.0368	0	0	1	0	0	0	0	0	0	1
18	88	235	1164	2627	36,276	29.2553	0	0	1	0	0	0	0	0	0	0
19	178	4	168	145,417	6367	40.1825	0	0	0	1	0	1	0	0	0	0
20	178	29	126	145,417	9959	73.4295	0	0	0	1	0	0	1	0	0	0
21	178	88	0	145,417	1508	63.7990	0	0	0	1	0	0	0	1	0	0
22	178	178	141,604	145,417	153,218	0.0000	0	0	0	1	0	0	0	0	1	0

Continued

Table 7.3 Dataset for estimating gravity model specifications—cont'd

ID	Origin FIPS code	Destination FIPS code	Migration	Origin total (O_i)	Destination total (D_j)	Distance	Origin I_1	Origin I_2	Origin I_3	Origin I_4	Origin I_5	Destination I_1	Destination I_2	Destination I_3	Destination I_4	Destination I_5
23	178	205	3227	145,417	19,029	18.8758	0	0	0	1	0	0	0	0	0	1
24	178	235	292	145,417	36,276	80.3585	0	0	0	1	0	0	0	0	0	0
25	205	4	651	25,345	6367	32.6467	0	0	0	0	1	1	0	0	0	0
26	205	29	39	25,345	9959	62.6176	0	0	0	0	1	0	1	0	0	0
27	205	88	9	25,345	1508	45.0368	0	0	0	0	1	0	0	1	0	0
28	205	178	9874	25,345	153,218	18.8758	0	0	0	0	1	0	0	0	1	0
29	205	205	14,699	25,345	19,029	0.0000	0	0	0	0	1	0	0	0	0	1
30	205	235	73	25,345	36,276	63.6952	0	0	0	0	1	0	0	0	0	0
31	235	4	21	37,040	6367	49.0638	0	0	0	0	0	1	0	0	0	0
32	235	29	2872	37,040	9959	32.8026	0	0	0	0	0	0	1	0	0	0
33	235	88	200	37,040	1508	29.2553	0	0	0	0	0	0	0	1	0	0
34	235	178	161	37,040	153,218	80.3585	0	0	0	0	0	0	0	0	1	0
35	235	205	12	37,040	19,029	63.6952	0	0	0	0	0	0	0	0	0	1
36	235	235	33,774	37,040	36,276	0.0000	0	0	0	0	0	0	0	0	0	0

Note: FIPS denotes Federal Information Processing Standards.

Table 7.4 SA variates

			Eigenvectors[a] of matrix $(I - 11^T/6)C(I - 11^T/6)$					SA-adjusted	
id	$LN(A_i)$	$LN(B_j)$	E_1	E_2	E_4	E_5	E_6	$LN(A_i)$	$LN(B_j)$
4	0.6563	0.7305	−0.33850	0.31953	0.00000	−0.53452	0.57527	−0.1003	−0.2293
29	−0.0265	1.3124	0.21592	0.69631	0.00000	0.00000	−0.54942	0.4403	0.8359
88	1.6888	0.7945	0.44315	−0.55033	0.00000	−0.53452	−0.21994	0.9991	0.3510
178	−1.2645	−0.1509	−0.42523	−0.23472	0.70711	0.26726	−0.16123	−0.4802	−0.4612
205	−0.2756	0.3337	−0.42523	−0.23472	−0.70711	0.26726	−0.16123	−0.0233	−0.7092
235	0	0	0.52988	0.00392	0.00000	0.53452	0.51656	0	0

[a]Because $\lambda_4 = \lambda_5 = -1$, eigenvectors $\mathbf{E}_4$ and $\mathbf{E}_5$ are not unique. The solutions appearing in this table are from R. SAS and Minitab produce different solutions, as other software packages may.

Note: $\mathbf{E}_3 = -0.408251$.

Source	DF	Sum of Squares	Mean Square	F Value	Approx Pr > F
Model	4	2.159E10	5.3964E9	8318.03	<.0001
Error	32	20760460	648764		
Uncorrected Total	36	2.161E10			

Parameter	Estimate	Approx Std Error	Approximate 95% Confidence Limits	
k	-0.7104	0.1700	-1.0566	-0.3642
a	0.2490	0.0701	0.1062	0.3919
b	0.8050	0.0649	0.6728	0.9372
g	-0.1158	0.00667	-0.1293	-0.1022

Approximate Correlation Matrix

	k	a	b	g
k	1.0000000	-0.4366458	0.2490249	-0.3612176
a	-0.4366458	1.0000000	-0.9800043	0.8095678
b	0.2490249	-0.9800043	1.0000000	-0.7919782
g	-0.3612176	0.8095678	-0.7919782	1.0000000

(A)

Criteria For Assessing Goodness Of Fit

Criterion	DF	Value	Value/DF
Deviance	32	10370.2772	324.0712
Scaled Deviance	32	32.0000	1.0000
Pearson Chi-Square	32	14906.6026	465.8313
Scaled Pearson X2	32	45.9979	1.4374
Log Likelihood		6914.4831	
Full Log Likelihood		-5306.1713	
AIC (smaller is better)		10620.3426	
AICC (smaller is better)		10621.6329	
BIC (smaller is better)		10626.6767	

Algorithm converged.

Analysis Of Maximum Likelihood Parameter Estimates

Parameter	DF	Estimate	Standard Error	Wald 95% Confidence Limits		Wald Chi-Square	Pr > ChiSq
Intercept	1	-0.9089	0.4756	-1.8411	0.0232	3.65	0.0560
lnoi	1	0.3808	0.0663	0.2509	0.5107	33.02	<.0001
lndj	1	0.6892	0.0609	0.5698	0.8085	128.10	<.0001
dist	1	-0.0911	0.0050	-0.1008	-0.0813	334.66	<.0001
Scale	0	18.0020	0.0000	18.0020	18.0020		

(B)

Criteria For Assessing Goodness Of Fit

Criterion	DF	Value	Value/DF
Deviance	24	6221.8892	259.2454
Scaled Deviance	24	24.0000	1.0000
Pearson Chi-Square	24	8076.1372	336.5057
Scaled Pearson X2	24	31.1525	1.2980
Log Likelihood		8651.4897	
Full Log Likelihood		-3231.9773	
AIC (smaller is better)		6487.9547	
AICC (smaller is better)		6501.5199	
BIC (smaller is better)		6506.9569	

Algorithm converged.

Analysis Of Maximum Likelihood Parameter Estimates

Parameter	DF	Estimate	Standard Error	Wald 95% Confidence Limits		Wald Chi-Square	Pr > ChiSq
Intercept	1	-10.5624	0.0857	-10.7303	-10.3944	15193.7	<.0001
dist	1	-0.0915	0.0049	-0.1011	-0.0820	351.18	<.0001
io1	1	0.6563	0.4352	-0.1967	1.5093	2.27	0.1316
io2	1	-0.0265	0.3631	-0.7381	0.6852	0.01	0.9419
io3	1	1.6888	0.4329	0.8402	2.5373	15.22	<.0001
io4	1	-1.2645	0.4675	-2.1808	-0.3482	7.32	0.0068
io5	1	-0.2756	0.4642	-1.1854	0.6343	0.35	0.5528
id1	1	0.7305	0.4581	-0.1674	1.6285	2.54	0.1108
id2	1	1.3124	0.3400	0.6460	1.9788	14.90	0.0001
id3	1	0.7945	0.5430	-0.2697	1.8587	2.14	0.1434
id4	1	-0.1509	0.4679	-1.0679	0.7660	0.10	0.7470
id5	1	0.3337	0.4723	-0.5920	1.2593	0.50	0.4799
Scale	0	16.1011	0.0000	16.1011	16.1011		

(C)

Analysis Of Maximum Likelihood Parameter Estimates

Parameter	DF	Estimate	Standard Error	Wald 95% Confidence Limits		Wald Chi-Square	Pr > ChiSq
Intercept	1	-9.6166	0.1860	-9.9812	-9.2521	2672.94	<.0001
dist	1	-0.1110	0.0021	-0.1151	-0.1068	2773.08	<.0001
io1	1	0.0328	0.3039	-0.5628	0.6284	0.01	0.9140
io2	1	0.5956	0.2745	0.0575	1.1336	4.71	0.0301
io3	1	1.1455	0.3758	0.4089	1.8821	9.29	0.0023
io4	1	-0.3588	0.2454	-0.8398	0.1222	2.14	0.1437
io5	1	0.1003	0.2826	-0.4537	0.6542	0.13	0.7228
id1	1	-0.2592	0.2624	-0.7734	0.2551	0.98	0.3233
id2	1	0.8447	0.2033	0.4462	1.2432	17.26	<.0001
id3	1	0.3495	0.2122	-0.0664	0.7654	2.71	0.0995
id4	1	-0.6214	0.2441	-1.0998	-0.1430	6.48	0.0109
id5	1	-0.6117	0.2595	-1.1204	-0.1031	5.56	0.0184
e3	1	1.1699	0.5880	0.0174	2.3224	3.96	0.0466
e5	1	-1.3159	0.4256	-2.1501	-0.4817	9.56	0.0020
e6	1	-2.5454	0.4686	-3.4638	-1.6269	29.51	<.0001
e7	1	-1.7497	0.4922	-2.7144	-0.7850	12.64	0.0004
e11	1	1.8673	0.3525	1.1765	2.5581	28.07	<.0001
e14	1	1.0714	0.2688	0.5446	1.5983	15.89	<.0001
e16	1	1.5156	0.6080	0.3239	2.7073	6.21	0.0127
e18	1	1.4223	0.5210	0.4011	2.4435	7.45	0.0063
e21	1	-2.5032	0.5361	-3.5540	-1.4524	21.80	<.0001
e23	1	0.9491	0.5810	-0.1897	2.0879	2.67	0.1024
e25	1	-1.3265	0.2883	-1.8915	-0.7614	21.17	<.0001
Scale	0	4.1501	0.0000	4.1501	4.1501		

(D)

Fig. 7.10 Estimation output for the gravity model specifications. (A) Eq. (7.2a)-normal. (B) Eq. (7.2b)-Poisson. (C) Eq. (7.3). (D) Eq. (7.5). Note: The SAS HPGENSELECT and R STEP function with the generalized linear model (GLM) do not choose exactly the same set of eigenvectors.

County	Origin ESF	Destination ESF
Aransas	−0.10032	−0.12267
Calhoun	0.10377	0.16463
Goliad	−0.10317	0.06811
Nuece	−0.17227	−0.22910
San Patricio	−0.03944	−0.30728
Victoria	−0.27062	0.02806

(A) (B) (C)

Fig. 7.11 Migration NA ESFs; values are directly proportional to a scale from *dark green* to *yellow* to *dark red.* (A) Origin NA ESF. (B) NA ESF values. (C) Destination NA ESF.

be customized for a researcher's own spatial interaction data analysis. All calculation results appearing in this section were verified with SAS.

7.6 Summary

This chapter presents a treatment of SA in spatial interaction data, a revisited topic in the literature highlighting Curry's (1972) insights about how SA and global distance decay intermingle. Many of the more recent empirical studies focus on journey-to-work data, with some focusing on migration. This chapter examines both phenomena, contributing to the formulation of a generalization about SA, NA, and spatial interaction. Continuing to build this foundation constitutes a worthwhile theme for fruitful future research, especially in terms of MESF methodology, which has helped to make the study of SA in spatial interaction data feasible. This chapter demonstrates that MESF furnishes insights about both SA and NA that are difficult, if not impossible, to glean from georeferenced data with other spatial analysis techniques. It also demonstrates that spatial interaction analyses should be accompanied by four maps, as well as effectively accommodate SA in order to obtain an unbiased distance decay parameter estimate. Tables 7.1 and 7.2 confirm that, once again, SA accounts for about 50% of excess Poisson variation and that other nongeographic sources of this overdispersion remain obscure. In addition, this chapter includes a toy example to illustrate various parts of the data analysis involved here.

Appendix 7.A A Corpus Christi toy spatial interaction dataset R code

```
## set a working directory
setwd("C:/IntroESF")

## load libraries and functions
library(rgdal)
library(spdep)
library(MASS)
source("functions.R")

cc.mig <- read.csv("CorpusChristi_MSA-migration_data.csv", header=
TRUE)
cc.mig$lnOi <- log(cc.mig$oi)
cc.mig$lnDj <- log(cc.mig$dj)
cc.mig$offs <- cc.mig$lnOi + cc.mig$lnDj
```

```
## nonlinear least squares
cc.nls <- nls(flows ~ exp(k + a*lnOi + b*lnDj + g*dist), data=cc.mig,
                      start=list(k=-1, a=1, b=1, g=-0.1))
summary(cc.nls)

cc.poi <- glm(flows ~ lnOi + lnDj + dist, family=quasipoisson,
data=cc.mig)

summary(cc.poi)
## doubly constraint model
cc.dc <- glm(flows ~ io1 + io2 + io3 + io4 + io5 +
                     id1 + id2 + id3 + id4 + id5 + dist,
                     offset=offs, family=quasipoisson, data=cc.mig)
summary(cc.dc)

fb <- grepl("io", names(coefficients(cc.dc)))
lnai <- c(coefficients(cc.dc)[fb], 0)
tb <- grepl("id", names(coefficients(cc.dc)))
lnbj <- c(coefficients(cc.dc)[tb], 0)

## read a shapefile
cc.poly <- readOGR(dsn=getwd(), layer="CorpusChristi_MSA")
cc.nb <- poly2nb(cc.poly)

## generate eigenvectors
n <- length(cc.nb)
B <- nb2mat(cc.nb, style="B")
M <- diag(n) - matrix(1, n, n)/n
MBM <- M %*% B %*% M
eig <- eigen(MBM, symmetric=T)
EV <- as.data.frame(eig$vectors)
colnames(EV) <- paste("EV",1:n, sep="")
EV <- as.data.frame(EV[, round(eig$values, 4) != 0])

## run stepwise regression for ESF
lm.ai.init <- lm(lnai ~ 1, data=EV)
lm.ai.full <- lm(lnai ~ ., data=EV)
ai.esf <- stepwise(lm.ai.full, lm.ai.init, sle=0.1, sls=0.1001)
# EV5 & EV1 are selected
summary(ai.esf)$r.square

lm.bj.init <- lm(lnbj ~ 1, data=EV)
lm.bj.full <- lm(lnbj ~ ., data=EV)
```

```
bj.esf <- stepwise(lm.bj.full, lm.bj.init, sle=0.1, sls=0.1001)
# no EV is selected

## ESF for network autocorrelation
cEV <- kronecker(as.matrix(EV), as.matrix(EV))
id.ord <- 1:NCOL(cEV)
colnames(cEV) <- paste("EV", id.ord, sep="")

cc.esf.df <- cbind(cc.mig[,c(4, 7:17, 20)], cEV)
glm.init <- glm(flows ~ io1 + io2 + io3 + io4 + io5 +
                        id1 + id2 + id3 + id4 + id5 + dist,
                        offset=offs, family=poisson, data=cc.esf.df)
glm.full <- glm(flows ~ . - offs,
                       offset=offs, family=poisson, data=cc.esf.df)
sel.scl <- sqrt(glm.init$deviance/glm.init$df.residual)
cc.dc.na <- step(glm.init, scope=list(upper=glm.full, lower=glm.
init), scale=sel.scl, trace=0)
summary(cc2.dc.na)
```

Appendix 7.B The functions.R code

The computer code is available at https://github.com/ywchun/sp_esf or http://bit.ly/sp_esf.

References

Anas, A. (1983). Discrete choice theory, information theory and the multinomial logit and gravity models. *Transportation Research Part B: Methodological*, *17*, 13–23.

Carrothers, G. (1956). An historical review of the gravity and potential concepts of human interaction. *Journal of the American Institute of Planners*, *22*, 94–102.

Cesario, F. (1977). A new interpretation of the "normalizing" or "balancing" factors of gravity type spatial models. *Journal of Socio-Economic Planning Sciences*, *11*, 131–136.

Chun, Y., & Griffith, D. (2011). Modeling network autocorrelation in space–time migration flow data: An eigenvector spatial filtering approach. *Annals of the AAG*, *101*, 523–536.

Chun, Y., & Griffith, D. (2013). *Spatial statistics and geostatistics*. Thousand Oaks, CA: SAGE.

Curry, L. (1972). Spatial analysis of gravity flows. *Regional Studies*, *6*, 131–147.

Farmer, C., & Oshan, T. (2017). In J. P. Wilson (Ed.), *Spatial Interaction*: The Geographic Information Science and Technology Body of Knowledge.

Fischer, M., & Griffith, D. (2008). Modeling spatial autocorrelation in spatial interaction data: A comparison of spatial econometric and spatial filtering specifications. *Journal of Regional Science*, *48*, 969–989.

Flowerdew, R., & Aitkin, M. (1982). A method of fitting the gravity model based on the Poisson distribution. *Journal of Regional Science*, *22*, 191–202.

Griffith, D. (2009). Spatial autocorrelation in spatial interaction: Complexity-to-simplicity in journey-to-work flows. In P. Nijkamp & A. Reggiani (Eds.), *Complexity and spatial networks: In search of simplicity* (pp. 221–237). Berlin: Springer-Verlag.

Griffith, D. (2011). Visualizing analytical spatial autocorrelation components latent in spatial interaction data: an eigenvector spatial filter approach. *Computers, Environment and Urban Systems*, *35*, 140–149.

Griffith, D., and Fischer, M. 2013. Constrained variants of the gravity model and spatial dependence: Model specification and estimation issues. Journal of Geographical Systems 15: 291–317; reprinted in R. Patuelli and G. Arbia (eds.), Spatial Econometric Interaction Modelling, Chapter 3, 37–66.

Griffith, D., Fischer, M., & LeSage, J. (2017). The spatial autocorrelation problem in spatial interaction modelling: A comparison of two common solutions. *Letters in Spatial and Resource Sciences*, *10*, 75–86.

LeSage, J., & Pace, K. (2008). Spatial econometric modeling of origin–destination flows. *Journal of Regional Science*, *48*(5), 941–968.

Sen, A., & Smith, T. (1995). *Gravity models of spatial interaction behavior*. Berlin: Springer.

Van Loan, C. (2000). The ubiquitous Kronecker product. *Journal of Computational and Applied Mathematics*, *123*, 85–100.

Wilson, A. (1967). A statistical theory of spatial distribution models. *Transportation Research*, *1*, 253–269.

CHAPTER 8

Space–time modeling

Chapter outline

Chapter 6 (e.g., Section 6.1) examines random effects (RE) terms, which decompose into spatially structured (SSRE), and spatially unstructured (SURE) components. That chapter concentrates on the SSRE component, which captures spatial autocorrelation (SA) latent in georeferenced data, noting that discussion about how to estimate the SURE component is a topic treated in this chapter. Eq. (6.1), which may be rewritten as follows, is the specification of interest for RE terms:

$$\mathbf{Y} = \mathbf{X}\boldsymbol{\beta}_{\mathbf{X}} + \boldsymbol{\xi} = \mathbf{X}\boldsymbol{\beta}_{\mathbf{X}} + (\mathbf{E}_{\mathrm{K}}\boldsymbol{\beta}_{\mathbf{E}} + \mathbf{Z}_{\mathrm{SURE}}) + \boldsymbol{\varepsilon}, \tag{8.1}$$

where $\boldsymbol{\xi}$ is a regression residual error term, $\boldsymbol{\varepsilon}$ is a $\mathrm{N}(\mathbf{0}, \mathbf{I}\sigma^2)$ standard independent and identically distributed (IID) error term, the eigenvector spatial filter (ESF) $\mathbf{E}_{\mathrm{K}}\boldsymbol{\beta}_{\mathbf{E}}$ is a SSRE (i.e., accounts for SA) term containing K eigenvectors, and $\mathbf{Z}_{\mathrm{SURE}}$ is a SURE (i.e., is void of SA) term; the specification for generalized linear models (GLMs) is similar. Estimating the RE term $(\mathbf{E}_{\mathrm{K}}\boldsymbol{\beta}_{\mathbf{E}} + \mathbf{Z}_{\mathrm{SURE}})$ and separating it from the residual error $\boldsymbol{\varepsilon}$ requires additional information. In Chapter 6, the SSRE component uses residual SA as the additional information to estimate it for cross-sectional georeferenced data. The SURE component requires, say, repeated measures furnished by a space–time series or prior distributions posited in a Bayesian

Spatial Regression Analysis Using Eigenvector Spatial Filtering
https://doi.org/10.1016/B978-0-12-815043-6.00008-2

analysis (e.g., Griffith, 2013). In practice, this additional information enables a RE term to be estimated, and then, in turn, the RE term is decomposed into a SSRE and a SURE component. This chapter focuses on the space–time series source of additional information in order to estimate a SURE component, which accounts for missing variables—sources of commonalities across time periods in addition to SA—among other misspecification issues (also see Griffith, 2008).

Conceptualizing the term ($\mathbf{E}_K\boldsymbol{\beta}_\mathbf{E} + \mathbf{Z}_{SURE}$) as a RE implies that it comprises independent variable values—not deliberately selected by an experimenter—that constitute a small subset of all possible values to which inferences are to be made, and/or its role is to account for heterogeneity/excess variance. One alternative is to conceptualize this term as a fixed effects one, which would require n indicator variables, one for each areal unit, and perhaps T indicator variables, one for each point in time. RE yield larger standard errors than are obtained with fixed effects[1] and are a theme of this chapter, in part because RE specifications address observations (i.e., areal units) obtained from other than a simple random sampling design.

Because of complexity, often only the intercept of a regression equation is implemented as a RE term (and is integrated out of a likelihood function with numerical methods such as empirical Bayes techniques). Doing so often requires invoking the following two important assumptions: (1) the intercept is a random outcome variable that is a function of a global mean value (the conventional constant intercept) plus a random error (deviations of local areal unit specific means from their global mean), and (2) the unobserved RE term is normally distributed and uncorrelated with each explanatory variable (which supports inference beyond the nonrandom sample analyzed). In other words, in its simplest form, a RE term is an intercept that is allowed to vary across areal units (where repeated observations are individual areal unit time series).

Treating a time series as repeated measures allows the uncovering of a temporal correlation-inducing common factor that spans time, which is a RE term (i.e., $\mathbf{E}_K\boldsymbol{\beta}_\mathbf{E} + \mathbf{Z}_{SURE}$). This term is areal unit specific and constant through time (i.e., it is the same for each point in time). Accordingly,

[1] A fixed effects specification establishes an equivalency between areal unit specific errors and fixed areal unit specific intercept terms, allowing the conditional mean of a response variable to vary across a geographic landscape. A RE specification treats these errors as uncorrelated random variables drawn from a population with zero mean and constant variance. This random sampling adds additional variance to RE standard errors.

Eq. (8.1) has a response variable, $\mathbf{Y}$, that is a nT-by-1 vector for n geographic locations and T points in time. The organization of this vector for statistical analysis purposes is n concatenated times series, each of length T. These time series are the repeated measures needed to estimate a SURE term. Each of the resulting n terms is constant across the T time periods and allows visualization of the component with a single map. The error term, $\boldsymbol{\varepsilon}$, represents a nT-by-1 vector of stochastic noise, and $\mathbf{X}_t\mathbf{b}_\mathbf{X} + (\mathbf{E}_K\mathbf{b}_\mathbf{E} + \hat{\mathbf{Z}}_{SURE})$ is the set of n predicted values for time period t (where $\mathbf{b}_\mathbf{X}$ denotes an estimated vector of regression coefficients for the covariates contained in matrix $\mathbf{X}_t$, $\mathbf{E}_K\mathbf{b}_\mathbf{E}$ is an estimated ESF, and $\hat{\mathbf{Z}}_{SURE}$ is an estimated SURE term). Table 8.1 presents the data organization described here.

8.1 Estimating a SURE term

This section summarizes RE results using the annual population density data for Texas counties during 2010–16.[2] These space–time data contain considerable redundant information because each year is based on the 2010 decennial census counts adjusted by officially recorded births, deaths, net domestic migration, and net international migration; a single principal components dimension describing these data accounts for 99.98% of the variance contained in the space–time data cube. These data exhibit moderate positive SA (PSA) [Moran coefficient (MC) = 0.52; Geary ratio (GR) = 0.44]. The conspicuous map pattern (Fig. 8.1A) closely reflects the geographic distribution of metropolitan areas across the state (Fig. 8.1B). The ensuing analyses are for a pure SA description of these data, with population counts treated as a Poisson random variable (RV) and density captured with a log-area offset variable.

The first step in an analysis of this type is to estimate the RE term. Here this is done with Poisson regression. The GLM has the expectation form

$$E[LN(pop_{it})] = \beta_0 + LN(area_i) + RE_i,$$

where E denotes the calculus of expectations, LN denotes natural logarithm, a log-link function is used, and RE $\sim N(\mathbf{0}, \sigma^2\mathbf{I})$. This n-by-1 vector is the same for each of the seven time periods (see Table 8.1). It has a mean of

[2] Elhorst (2010) considers T = 7 small. Small T may result in the time dimension failing to furnish enough information for consistent estimation of temporal autocorrelation effects. Lee and Yu (2012) emphasize that this situation can be remedied if n is sufficiently large, hence compensating for a small T. They also find that RE specifications tend to be preferable to fixed effects specifications.

Table 8.1 A partial portion of the data for two selected Texas counties

County	Year	Population	LN(area)	$b_0 \times 1$	RE	RE standard error	SSRE	SURE
Menard	2016	2123	5.4537	0.4179	−2.1836	0.1125	−1.1758	−1.0077
Menard	2015	2159	5.4537	0.4179	−2.1836	0.1125	−1.1758	−1.0077
Menard	2014	2154	5.4537	0.4179	−2.1836	0.1125	−1.1758	−1.0077
Menard	2013	2152	5.4537	0.4179	−2.1836	0.1125	−1.1758	−1.0077
Menard	2012	2227	5.4537	0.4179	−2.1836	0.1125	−1.1758	−1.0077
Menard	2011	2223	5.4537	0.4179	−2.1836	0.1125	−1.1758	−1.0077
Menard	2010	2236	5.4537	0.4179	−2.1836	0.1125	−1.1758	−1.0077
Hale	2016	34,263	5.5615	0.4179	0.4948	0.1122	0.2795	0.2153
Hale	2015	34,201	5.5615	0.4179	0.4948	0.1122	0.2795	0.2153
Hale	2014	34,537	5.5615	0.4179	0.4948	0.1122	0.2795	0.2153
Hale	2013	35,751	5.5615	0.4179	0.4948	0.1122	0.2795	0.2153
Hale	2012	36,284	5.5615	0.4179	0.4948	0.1122	0.2795	0.2153
Hale	2011	36,387	5.5615	0.4179	0.4948	0.1122	0.2795	0.2153

Note: LN denotes natural logarithm.

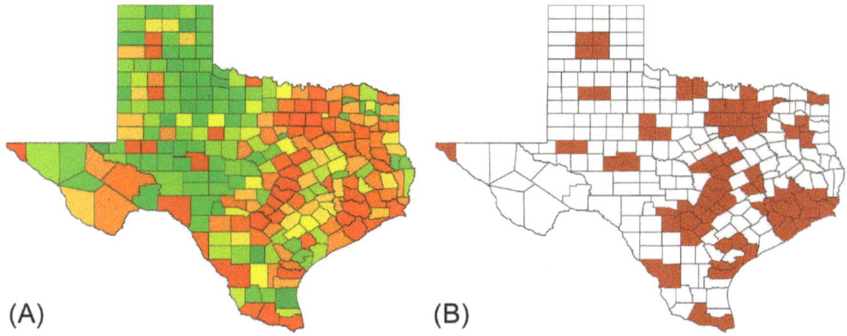

Fig. 8.1 Texas population features by county. (A) Annual 2010–16 log-population density principal components scores (which are directly proportional to a scale going from *dark green* to *yellow* to *dark red*). (B) *Red denotes* Texas census metropolitan area counties.

0.00070, and a Shapiro–Wilk normality diagnostic statistic probability of 0.71 (i.e., fail to reject the null hypothesis that the RE term conforms to a normal curve). The second step in this analysis is to estimate the SSRE and the SURE components. This decomposition can be achieved with a linear regression analysis in which the response variable is the RE term, and the covariates are the eigenvectors of the modified spatial weights matrix (SWM; see Section 2.1.3). The fitted values constitute the SSRE component, and the residuals constitute the SURE component; the intercept should be zero (more precisely, 0.00070 here). Table 8.2 reveals that the SSRE is a linear combination of 29 modified SWM eigenvectors [24 portraying PSA and 5 portraying negative SA (NSA); the respective candidate set sizes are 65 and 89], accounting for roughly two-thirds of the variance in the RE term, and captures moderate to strong PSA (Fig. 8.2A), whereas the SURE accounts for roughly one-third of the variance in the RE term and contains only a trace amount of SA (Fig. 8.2B). Inclusion of the RE term in a

Table 8.2 Summary statistics for the RE term

Feature	SSRE	SURE
% of variance accounted	64	36
MC	0.81	−0.00
GR	0.24	0.91

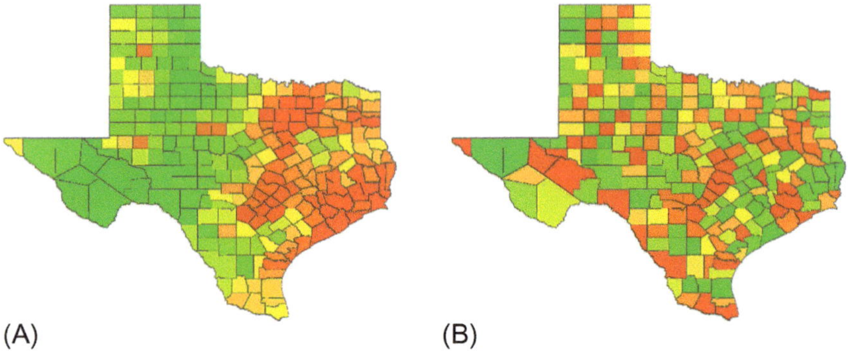

Fig. 8.2 The 2010–16 population density RE term (values are directly proportional to a scale going from *dark green* to *yellow* to *dark red*). (A) The SSRE component. (B) The SURE component.

Table 8.3 Quasi-likelihood estimates of the overdispersion parameter φ for a Poisson RV

Year	2010	2011	2012	2013	2014	2015	2016
Without RE	581.62	587.54	593.98	599.84	606.43	612.91	618.51
With RE	7.98	6.06	3.92	1.40	2.86	6.07	9.18

Note: for each year, the NB overdispersion parameter, $\hat{\eta}$, is <0.001.

spatial analysis accounts for extra-Poisson variation in the population counts data (see Table 8.3), resulting in an avoidance of switching to a negative binomial (NB) probability model (for a discussion about the possible inappropriateness of this substitution for georeferenced data containing SA, see Section 5.4.2). The central point in time, year 2013, highlights that the RE best fits the data in the center of the time series; here is where it reduces the quasi-likelihood estimate of the overdispersion parameter φ to its lowest value.

The trend lines based on the RE term are similar for each year (Fig. 8.3A–C), their plots ordered by year, with the 2013 trend line in the center. The SSRE term deviates (Fig. 8.3B) and the SURE term dramatically deviates (Fig. 8.3C) from the trend lines. The multiplication of these two terms yields the RE plot (Fig. 8.3A), indicating that ESF values either inflate $[\exp(ESF_i) > 1]$, leave the same $[\exp(ESF_i) \approx 1]$, or deflate $[\exp(ESF_i) < 1]$ population density values, depending upon the SA in the local region of areal unit i.

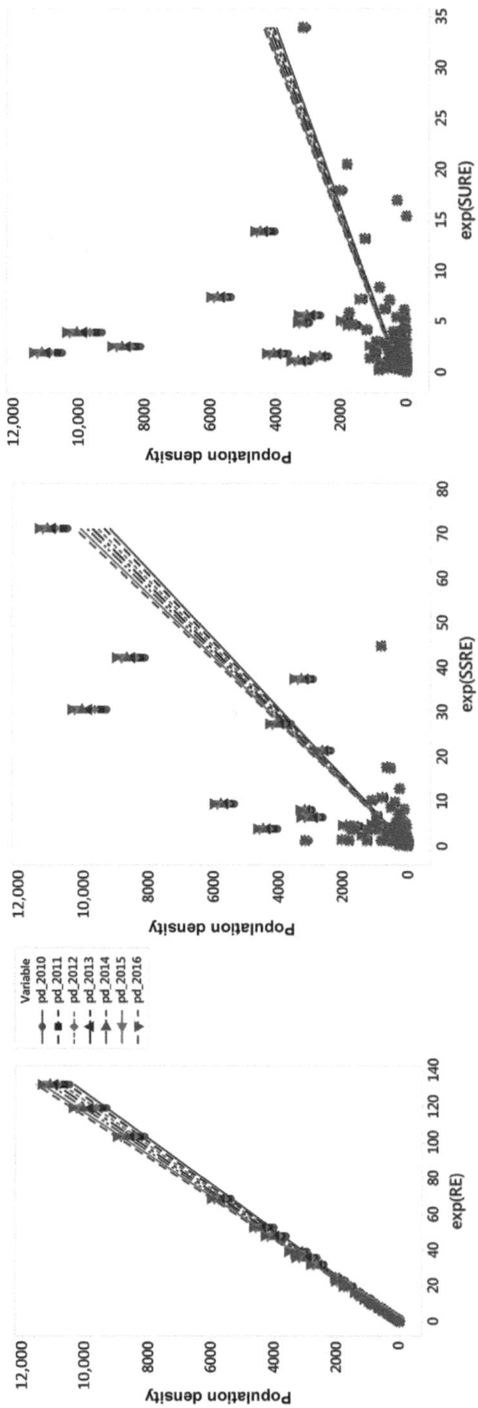

Fig. 8.3 Trend lines for population density versus the exponentiated RE term/component.

8.1.1 A RE term estimation sensitivity analysis

One relevant question about a RE term is whether its estimation is robust with regard to the space–time data series employed as repeated measures. This question is addressed here in terms of the following three geographic distributions of Texas population density estimation scenarios: (1) subsets of maps equally spaced in time, (2) pairs of maps increasingly separated in time, and (3) a Bayesian map analysis based on principal components scores. Fig. 8.4A portrays the first scenario and involves three subsets of maps. Fig. 8.4B portrays the second scenario and involves six pairs of maps. Fig. 8.4C portrays the third scenario, for which the Bayesian RE term was calculated using an ICAR (i.e., improper intrinsic multivariate conditional autoregressive) auto-normal probability model specification implemented in GeoBUGS (Thomas, Best, Lunn, Arnold, & Spiegelhalter, 2004) of WinBUGS. Fig. 8.4 illustrates that the RE terms estimated in these various ways essentially are identical.

8.1.2 Prediction based on an estimated RE term

Prediction error—the average difference between pairs of predicted and observed values—is a useful way to assess the reliability of a model specification. Two features are of interest here. One concerns stability of the ESF across cross-validation subsamples. The other concerns the relative size of prediction error vis-à-vis the fitted model mean squared error (MSE). Given results summarized in the preceding section, analyses for this section employed only 2010 and 2016 data, and missing data are assumed to be an entire county-specific time series.

Imputed RE values are zero in this case. The average R^2 for the leave-one-out analysis is 0.645 ($s_{R^2}=0.007$), which is consistent with the complete dataset value reported in Table 8.2. Table 8.4 indicates that the selected eigenvectors are reasonably stable, almost always being selected in a cross-validation case. On occasion, ESF construction selected from among 22 other eigenvectors contained in the remainder candidate set of 125 eigenvectors, some for only one cross-validation case (e.g., $\mathbf{E}_5$), and others for as many of 35 cross-validation cases ($\mathbf{E}_{36}$).

For 2010 and 2016, respectively, the bivariate regression equation for observed and predicted population density yields a MSE of 649.06 and 1020.17 based on the complete RE term, and 1,067,815 and 1,234,352 based on only the SSRE term. An ESF capturing the SSRE component furnishes at least part of the RE term for imputation purposes and, respectively,

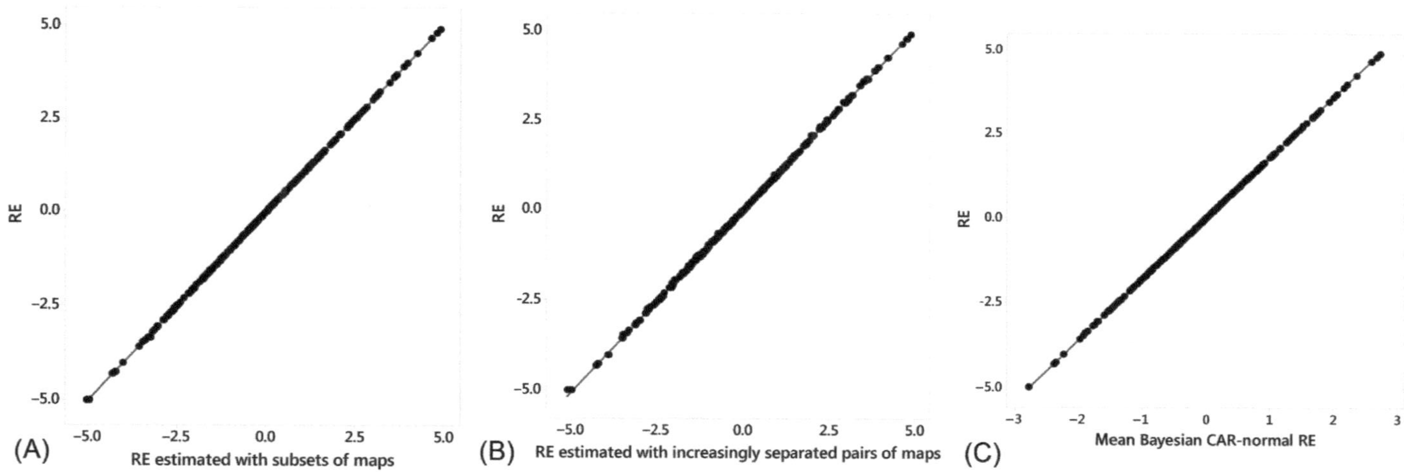

Fig. 8.4 Scatterplot comparisons of the approximated RE with the complete RE estimate. (A) Equally spaced in time subsets of maps. (B) Increasingly separated in time pairs of maps. (C) Bayesian map analysis with a normal prior.

Table 8.4 Cross-validation results for the complete dataset ESF eigenvectors

	Complete dataset stepwise regression results		
Eigenvector	**Significance probability**	**Partial R^2**	**Cross-validation frequency of selection**
E_1	<0.0001	0.0519	254
E_2	<0.0001	0.0056	254
E_3	<0.0001	0.0090	254
E_4	<0.0001	0.0531	254
E_8	<0.0001	0.0050	254
E_9	0.0001	0.0059	254
E_{10}	0.0625	0.0055	250
E_{11}	0.0180	0.0125	254
E_{13}	<0.0001	0.0139	254
E_{14}	0.0767	0.0620	240
E_{15}	0.0562	0.0061	250
E_{18}	0.0035	0.0052	254
E_{21}	0.0515	0.0173	253
E_{23}	0.0011	0.0083	254
E_{25}	<0.0001	0.0057	254
E_{27}	0.0764	0.0484	242
E_{28}	0.0042	0.0050	254
E_{35}	0.0040	0.0133	254
E_{38}	0.0623	0.1489	249
E_{42}	0.0856	0.0135	227
E_{43}	0.0402	0.0056	254
E_{52}	0.0043	0.0276	254
E_{56}	<0.0001	0.0047	254
E_{61}	0.0544	0.0068	250
E_{166}	0.0635	0.0133	251
E_{167}	0.0056	0.0252	254
E_{220}	0.0731	0.0060	249
E_{235}	0.0236	0.0340	254
E_{248}	0.0590	0.0238	251

yields a mean squared prediction error (MSPE) of 957,743 and 1,219,771. The imputed values display the typical funnel scatterplot for both years (Fig. 8.5). These results imply that the SURE term plays an important role in describing georeferenced data, even though here the SSRE accounts for roughly two-thirds of the geographic variation in the RE term. They also imply that, for a nonnormal RV, the MSPE compares favorably to the MSE for fitted values when both are measured by ordinary least squares (OLS) estimated bivariate linear regression.

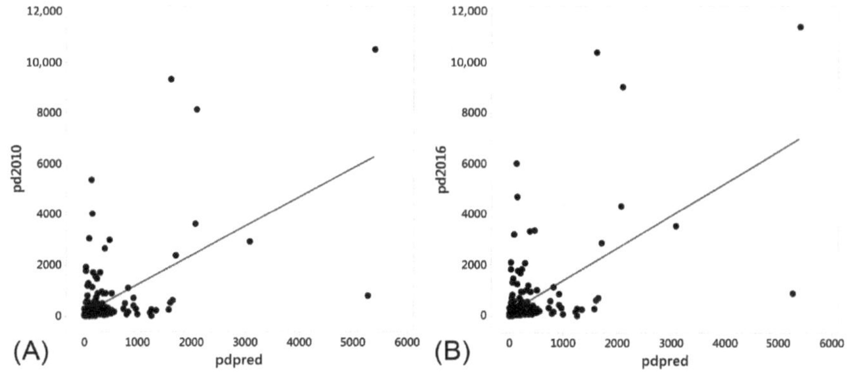

Fig. 8.5 Scatterplots of the observed Texas population density values versus their imputed values based on only the SSRE component estimate and county mean. (A) For 2010. (B) For 2016.

8.2 Space–time data structures: Eigenvector space–time filters

The preceding discussion casts space–time dependence in terms of a common factor spanning time that has a SSRE component. Various space–time data structures exist for its competing autoregressive types of conceptualizations, which relate more to fixed effects varieties of specifications. Fig. 8.6

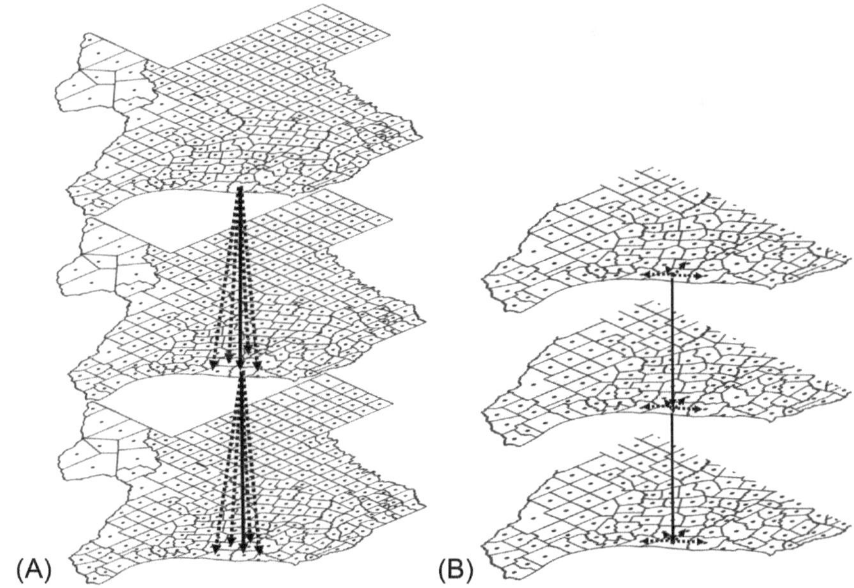

Fig. 8.6 Space–time data structures for Texas counties. (A) Time-lagged spatial dependence. (B) Contemporaneous spatial dependence.

presents two popular versions, namely, the lagged and the contemporaneous spatial dependence perspectives. In each case, the value of some attribute variable at location i during time t, $y_{i,t}$, is a function of the preceding time period's value, $y_{i,t-1}$, for that location; this is a standard first-order time series structure. Time-lagged spatial dependence specifies $y_{i,t}$ also as a function of the neighboring values of location i during time t–1, $\mathbf{WY}_{t-1}$: the SWM applies to the previous time period (i.e., a lagged spatial structure; Fig. 8.6A). In contrast, contemporaneous spatial dependence specifies $y_{i,t}$ also as a function of the neighboring values of location i during time t, $\mathbf{WY}_t$: the SWM applies to the same time period (i.e., a contemporaneous spatial structure; Fig. 8.6B). Cliff and Ord (1981) and Griffith (1981) extend the MC to these space–time structures in terms of a space–time MC, namely, MC_{ST}.

In contrast to the purely common factor specification examined in Section 8.1, which involves the two-step procedure that first estimates a RE term and then decomposes it into a SSRE and a SURE component, here the eigenvector space–time filter (ESTF) is estimated first and then the SURE is estimated conditional on the estimated ESTF.

The next two sections summarize these two space–time conceptualizations, emphasizing how they allow Moran eigenvector spatial filtering (MESF) to be extended to space–time data to construct ESTFs. Doing so exploits an integration of the MC and the Durbin and Watson (1951) statistic.[3] The resulting adjusted space–time weights matrix (STWM) renders the modified fundamental MESF theorem (see Section 2.1.3):

> *The first eigenvector, say $\mathbf{E}_1$, is the set of real numbers that has the largest space–time MC achievable by any set for the three-dimensional articulation defined by the space–time connectivity matrix $\mathbf{A}$; the second eigenvector is the set of real numbers that has the second largest achievable space–time MC by any set that is orthogonal and uncorrelated with $\mathbf{E}_1$; the third eigenvector is the third such set of real numbers; and so on through $\mathbf{E}_{nT}$, the set of real numbers that has the largest negative space–time MC achievable by any set that is orthogonal and uncorrelated with the preceding (nT − 1) eigenvectors (Griffith, 2012).*

[3] The Durbin–Watson statistic may be written in matrix notation as

$$\frac{\mathbf{Y}^T\left[\mathbf{I}-\mathbf{X}(\mathbf{X}^T\mathbf{X})^{-1}\mathbf{X}^T\right]\mathbf{A}\left[\mathbf{I}-\mathbf{X}(\mathbf{X}^T\mathbf{X})^{-1}\mathbf{X}^T\right]\mathbf{Y}}{\mathbf{Y}^T\left[\mathbf{I}-\mathbf{X}(\mathbf{X}^T\mathbf{X})^{-1}\mathbf{X}^T\right]\mathbf{Y}}, \text{ where matrix } \mathbf{A} \text{ is } \begin{pmatrix} 1 & -1 & 0 & \cdots & 0 & 0 & 0 \\ -1 & 2 & -1 & \cdots & 0 & 0 & 0 \\ 0 & -1 & 2 & \cdots & 0 & 0 & 0 \\ 0 & 0 & -1 & \cdots & 0 & 0 & 0 \\ \vdots & \vdots & \vdots & \ddots & \vdots & \vdots & \vdots \\ 0 & 0 & 0 & \cdots & -1 & 2 & -1 \\ 0 & 0 & 0 & \cdots & 0 & -1 & 1 \end{pmatrix},$$

with its eigenvectors having elements $e_{ij} = \frac{1}{\sqrt{T+1}}\text{SIN}\left(\frac{ij\pi}{T+1}\right)$.

The two components referred to in this theorem, space and time, may have their parts of the STWM denoted by the n-by-n matrix $\mathbf{C}_S$ for space, and the T-by-T matrix $\mathbf{C}_T$ for time. The former is the SWM appearing in other chapters of this book. The latter, mirroring the Durban-Watson statistic, has all zero entries except for the upper and lower off-diagonal cells, whose entries are one. The following projection matrix now is nT-by-nT, as is its identity matrix $\mathbf{I}$, with vector $\mathbf{1}$ now being nT-by-1: $\mathbf{M} = \left(\mathbf{I} - \frac{\mathbf{1}\mathbf{1}^T}{nT}\right)$.

8.2.1 The space–time lagged spatial structure specification: Results for Texas population density

The matrix specification of the STWM for this conceptualization may be written as

$$\mathbf{M}[\mathbf{C}_T \otimes (\mathbf{C}_S + \mathbf{I}_S)]\mathbf{M},$$

(Griffith & Heuvelink, 2012) where ⊗ denotes Kronecker product. Because of this Kronecker product form, eigenfunctions need to be calculated only for matrices $\mathbf{C}_S$ and $\mathbf{C}_T$; these eigenfunctions can be combined with the matrix operator ⊗ to establish the space–time eigenfunctions needed to construct ESTFs.

Texas space–time population density exhibits considerable overdispersion, implying that its appropriate probability model may be a NB one. Effectively and efficiently accounting for SA in a mean response specification supports this connection. Meanwhile, the lagged structure removes the year 2010 from the data analysis, because it serves as the previous year's data for 2011. The adjusted STWM yields 204 eigenvectors with a standardized space–time MC ≥ 0.25, which constitutes the candidate set of eigenvectors for this specification. The construction of the ESTF selected 84 of these vectors; this ESTF accounts for roughly 70% of the space–time variance in population density across Texas during 2010–16. This ESTF reduces the overdispersion parameter (η) from 2.4969 to 0.7810, implying that space–time SA accounts for roughly 70% of the excess variance in these population density data. This space–time specification outperforms the preceding common factor RE specification in terms of both bivariate linear regression R^2 values for predicting population density from a spatially structured component (i.e., SSRE) and reduction in excess variance (Table 8.5).

Fig. 8.7 portrays the six individual geographic parts of the lagged ESTF. Their differences are subtle, and each highlights what essentially are the major urban regions of Texas (i.e., Austin–San Antonio, Brownsville–Corpus Christi, Dallas–Fort Worth, El Paso, and Houston).

Table 8.5 Summary statistics for space–time specifications

		Lagged			Contemporaneous			Common factor RE (Section 8.1.2)		
Year	η	ESTF-R^2	R^2	η	ESTF-R^2	R^2	η	SSRE-R^2	R^2	η
2010	2.4415				0.8288	0.9513	0.6871	0.6871	0.9995	1.0656
2011	2.4560	0.7884	0.9903	0.7307	0.8028	0.9833	0.6078	0.6002	0.9998	1.0728
2012	2.4742	0.7931	0.9979	0.7140	0.7663	0.9834	0.6068	0.6001	0.9999	1.0828
2013	2.4882	0.7860	0.9648	0.7479	0.7265	0.9596	0.6254	0.5993	>0.9999	1.0884
2014	2.5056	0.7909	0.9973	0.7200	0.7679	0.9833	0.6033	0.5987	>0.9999	1.0967
2015	2.5191	0.7875	0.9929	0.7437	0.8061	0.9879	0.5976	0.5990	0.9998	1.0996
2016	2.5346	0.7275	0.8948	0.8111	0.8292	0.9554	0.6745	0.6001	0.9994	1.1028
pooled	2.4969	0.6961	0.8855	0.7810	0.7240	0.9178	0.6387	0.5988	0.9997	1.0873

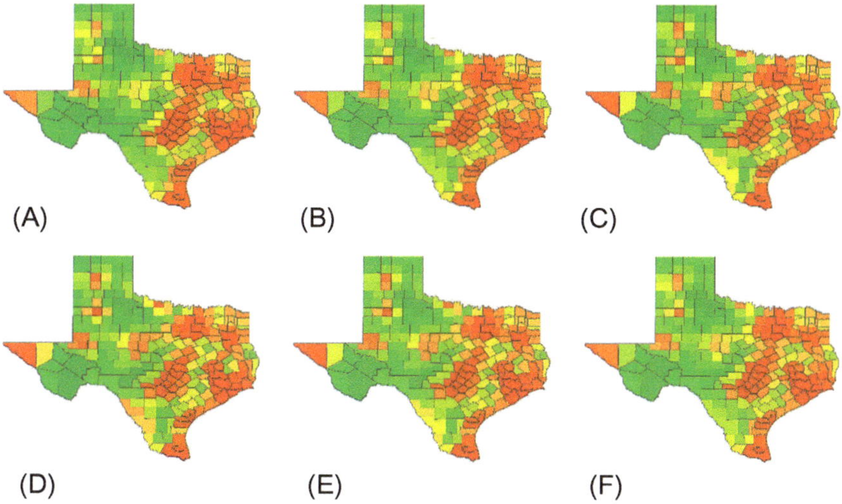

Fig. 8.7 Individual map components of the lagged ESTF; the values of *dark green* to *yellow* to *dark red* are directly proportional to the ESTF values. (A) 2016. (B) 2015. (C) 2014. (D) 2013. (E) 2012. (F) 2011.

Time series eigenvectors link together and differentially weight each of these six maps.

8.2.2 The space–time contemporaneous spatial structure specification: Results for Texas population density

The matrix specification of the STWM for this conceptualization may be written as

$$\mathbf{M}[\mathbf{I}_T \otimes \mathbf{C}_S + \mathbf{C}_T \otimes \mathbf{I}_S)]\mathbf{M}$$

(Griffith & Heuvelink, 2012). This contemporaneous structure retains year 2010 data in the space–time analysis. The adjusted STWM for this specification yields 414 eigenvectors with a standardized space–time MC ≥ 0.25, which constitutes its candidate set of eigenvectors. The construction of the ESTF selected 121 of these vectors; this ESTF accounts for roughly 72% of the space–time variance in population density across Texas during 2010–16. This ESTF reduces the overdispersion parameter (η) from 2.4969 to 0.6387, implying that space–time SA accounts for roughly 72% of the excess variance in these population density data. Overall, this space–time specification outperforms both of the preceding specifications (lagged and RE) in terms of both bivariate linear regression R^2 values for predicting population density from a spatial structure component, and reduction in excess variance (Table 8.5); the ESTF lagged specification results have slightly higher R^2 values in a few cases, and the common factor RE (SSRE + SURE) specification has higher R^2 values.

Fig. 8.8 portrays the pairwise scatterplots for the lagged and contemporaneous ESTFs[4] and demonstrates that the two conceptualizations render somewhat different eigenfunction descriptions. A comparison of Figs. 8.8A and B reveal that their composite and individual year scatterplots are very similar.

Fig. 8.9 portrays the six individual geographic parts of the contemporaneous ESTF. Their differences are subtle, and, again, each highlights what essentially are the major urban regions of Texas (i.e., Austin–San Antonio, Brownsville–Corpus Christi, Dallas–Fort Worth, El Paso, and Houston). As before, time series eigenvectors link together and differentially weight each of these six maps. Relatively little difference exists between Figs. 8.7 and 8.9 counterparts.

8.2.3 ESTF prediction

The preceding discussion (Section 8.1.2) indicates that a RE value cannot be estimated for a missing areal unit time series. However, it can be estimated with a partial time series (see Section 8.1.1). Meanwhile, the contemporaneous ESTF appears to outperform the lagged ESTF. For this section, sets of values have been randomly suppressed, a contemporaneous ESTF specification has been employed, and a RE term has been estimated. Because $\mathbf{C}_S$ and $\mathbf{C}_T$ are known, the eigenvectors always are known for any missing $y_{i,t}$ values. Different percentages of missing values—randomly sampled without replacement—have been imputed to calculate the MSPE; calculation of the MSE used the sampled values.

Table 8.6 presents a survey of prediction error; because of the numerical intensity involved, it is based on only 100 samples, making it exploratory in nature. The MSPE gains considerable precision by including a SURE term. Fig. 8.10 reveals a funnel-shaped scatter of points (typical of a Poisson RV), that the confidence interval shrinks and becomes nearly invisible with increasing missing values sample size, that many of the predicted values fall outside of their corresponding prediction interval, and that the spread of predicted values increases with an increase in percentage of imputations; some of these results are well known.

[4] Only six years of the contemporaneous ESTF appear here because the lagged ESTF loses 2010. Both ESTFs display symmetry in time, divulging the missing year portion of the lagged ESTF, making it available for inclusion.

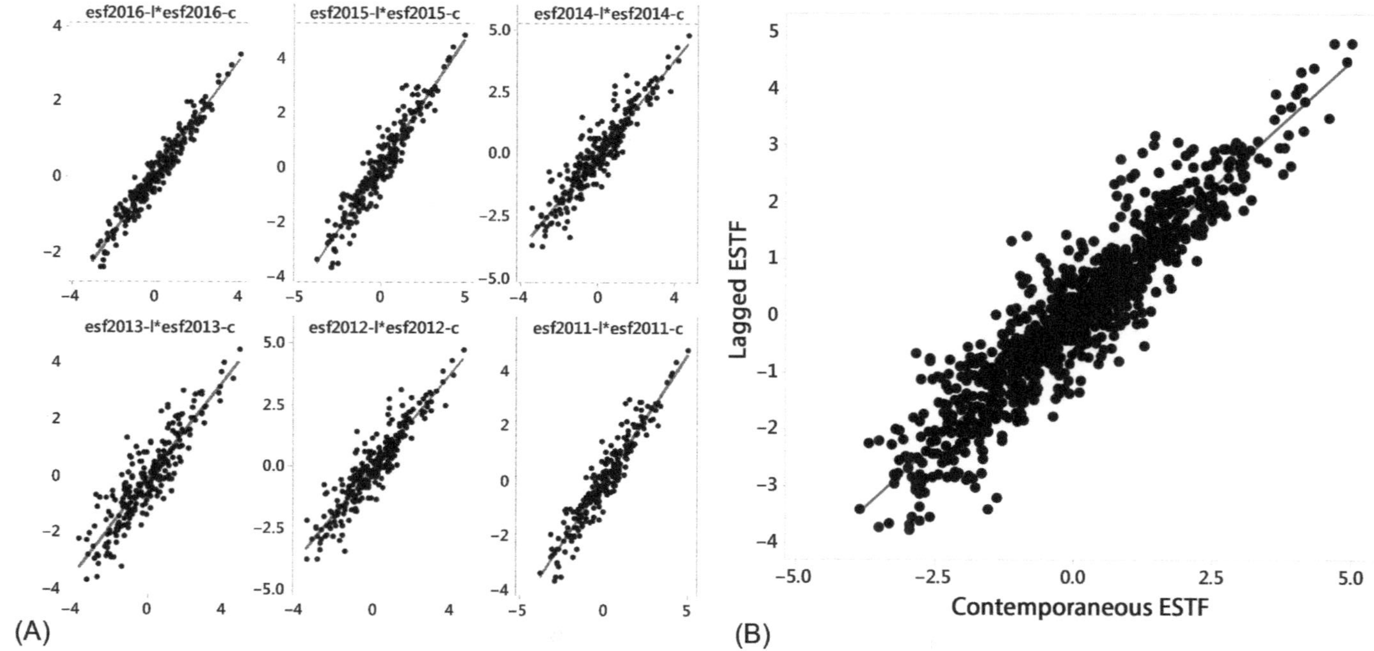

Fig. 8.8 Scatterplots of the ESTFs. (A) Individual year ESTFs (equivalent to ESFs). (B) Composite ESTFs.

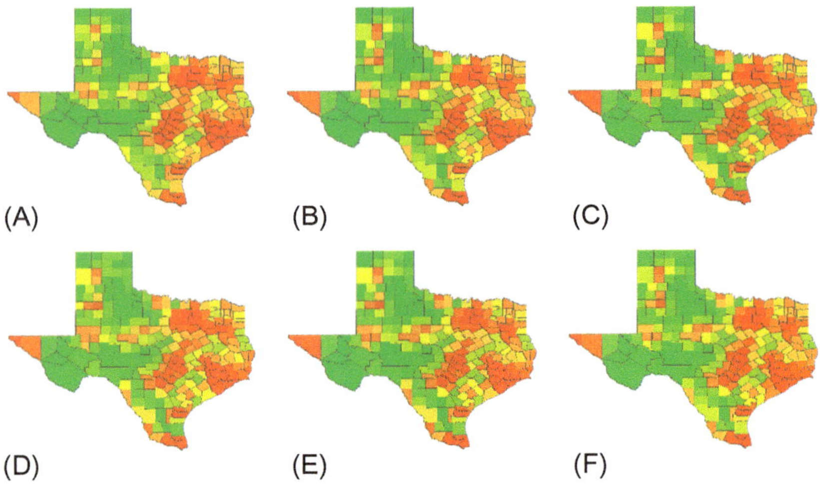

Fig. 8.9 Individual map components of the contemporaneous ESTF; the values of *dark green* to *yellow* to *dark red* are directly proportional to the ESTF values. (A) 2016. (B) 2015. (C) 2014. (D) 2013. (E) 2012. (F) 2011.

Table 8.6 MS(P)E simulated sampling results, 100 replications (n = 1778; MSE = 129,511)

% suppressed	# of imputations	MSE_{ESTF}	MSE_{ESTF}/ MSE	$MSPE_{RE}$	$MSPE_{RE}$/ MSE
0.1	2	85,589	0.66	333,696	2.58
10	178	664,842	5.13	305,789	2.36
20	356	790,393	6.10	384,318	2.97
30	533	823,196	6.36	400,672	3.09
40	711	916,198	7.07	649,283	5.01
50	889	1,296,848	10.01	882,670	6.82

Note: Subscript ESTF denotes that only an ESTF was used to calculate predicted values; subscript RE denotes that both an ESTF and a SURE value were used to calculate imputations.

8.3 A toy example: Exemplifying the necessary data structures

Because the data structure required for this type of analysis is complex and requires several reorganizations during analysis, this section presents a simple numerical example for pedagogic purposes. This example is based on the six Texas counties constituting the 2010 Corpus Christi metropolitan statistical area (MSA). The illustrative attribute is Box–Cox transformed population density for 2010, 2013, and 2016. The resulting toy example has n = 6 and T = 3 (nT = 18).

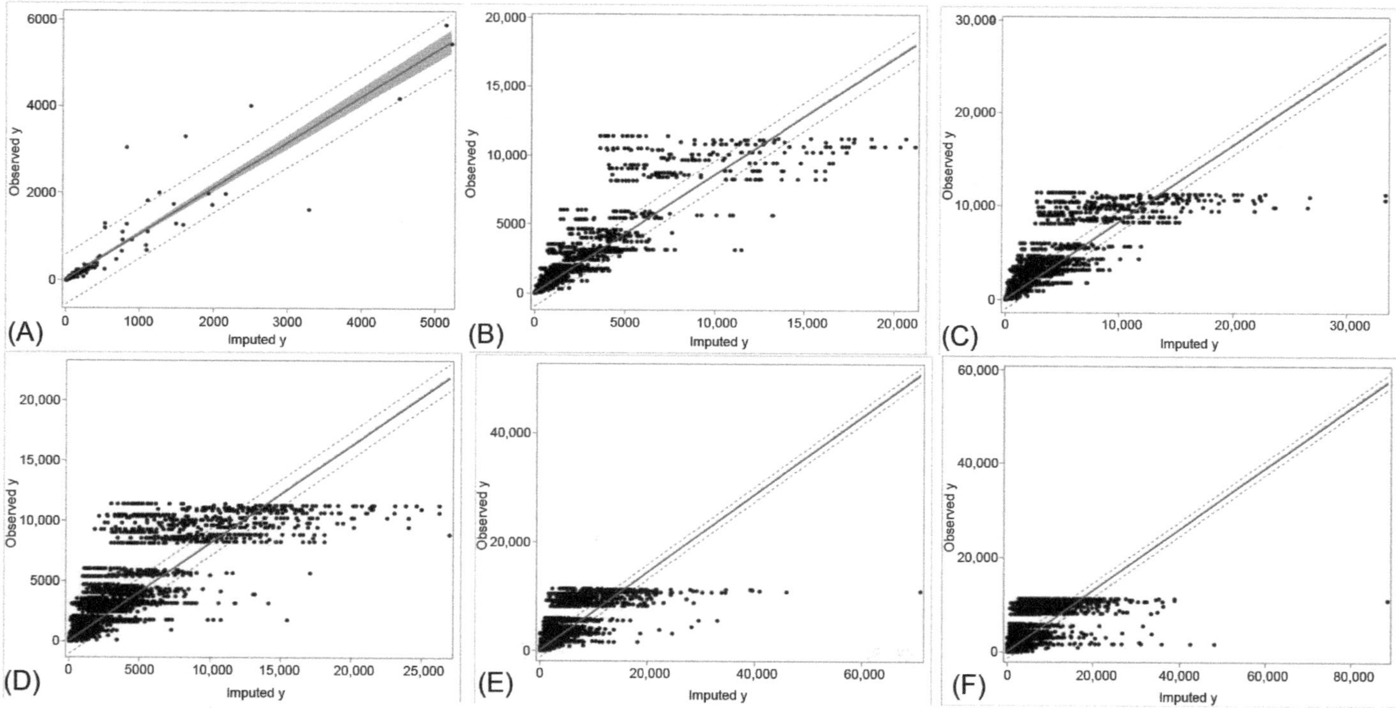

Fig. 8.10 Scatterplots for observed versus different percentages of imputed population densities, with superimposed confidence and prediction intervals. (A) 0.1%. (B) 10%. (C) 20%. (D) 30%. (E) 40%. (F) 50%.

Fig. 8.11 portrays the geographic distribution of LN (population density) for the three time periods using the same threshold classification values for each year. Fig. 8.12 displays the county numbering used for identification purposes, and the rook's adjacency-based binary SWM. Table 8.7 tabulates the original data, and the data reorganization needed to estimate a RE[5] term; this reorganization creates six individual county time series. Table 8.8 reports the output data, highlighting that (1) the intercept is constant across counties and time periods, (2) the RE is constant across time periods for each county, and (3) the residuals are space–time specific. With n = 6, no SWM

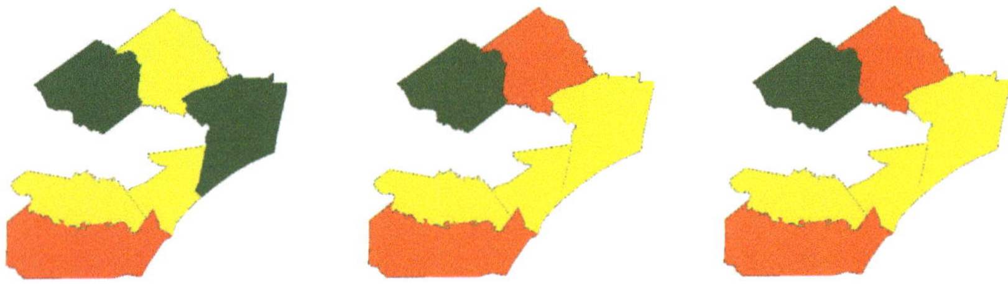

(A) (B) (C)

Fig. 8.11 Box–Cox transformed population density, LN(population/area − 0.1), for the Corpus Christi MSA; *green denotes* relatively low, *yellow denotes* moderate, and *red denotes* relatively high density. (A) 2010. (B) 2013. (C) 2016.

(A)

	1	2	3	4	5	6
1	**0**	**1**	**0**	**1**	**1**	**0**
2	**1**	**0**	**0**	**0**	**0**	**1**
3	**0**	**0**	**0**	**0**	**0**	**1**
4	**1**	**0**	**0**	**0**	**1**	**0**
5	**1**	**0**	**0**	**1**	**0**	**0**
6	**0**	**1**	**1**	**0**	**0**	**0**

(B)

Fig. 8.12 The Corpus Christi MSA spatial structure. (A) Areal unit names. (B) SWM, rook's adjacency definition.

[5] Because n = 6 is so small, estimation of the RE term requires judiciously selected starting values for its mean and variance. Otherwise convergence is problematic. This selection may involve some trial and error.

Table 8.7 Corpus Christi MSA attribute data structure

Original data structure					Reorganized data structure for estimating SSRE and SURE				
Year	**ID**	**Population**	**Area**	**County**	**Year**	**ID**	**Population**	**Area**	**County**
2016	1	23,199	0.125542	Aransas	2016	1	23,199	0.125542	Aransas
2016	2	21,327	0.246358	Calhoun	2013	1	24,124	0.125542	Aransas
2016	3	7223	0.205430	Goliad	2010	1	25,721	0.125542	Aransas
2016	4	340,314	0.276254	Nueces	2016	2	21,327	0.246358	Calhoun
2016	5	64,484	0.168225	San Patricio	2013	2	21,759	0.246358	Calhoun
2016	6	86,861	0.212752	Victoria	2010	2	21,965	0.246358	Calhoun
2013	1	24,124	0.125542	Aransas	2016	3	7223	0.205430	Goliad
2013	2	21,759	0.246358	Calhoun	2013	3	7446	0.205430	Goliad
2013	3	7446	0.205430	Goliad	2010	3	7517	0.205430	Goliad
2013	4	352,728	0.276254	Nueces	2016	4	340,314	0.276254	Nueces
2013	5	66,274	0.168225	San Patricio	2013	4	352,728	0.276254	Nueces
2013	6	90,075	0.212752	Victoria	2010	4	361,350	0.276254	Nueces
2010	1	25,721	0.125542	Aransas	2016	5	64,484	0.168225	San Patricio
2010	2	21,965	0.246358	Calhoun	2013	5	66,274	0.168225	San Patricio
2010	3	7517	0.205430	Goliad	2010	5	67,655	0.168225	San Patricio
2010	4	361,350	0.276254	Nueces	2016	6	86,861	0.212752	Victoria
2010	5	67,655	0.168225	San Patricio	2013	6	90,075	0.212752	Victoria
2010	6	92,467	0.212752	Victoria	2010	6	92,467	0.212752	Victoria

Note: Area is multiplied by 100,000 for analysis purposes.

Table 8.8 The data structure for the RE and population density residuals

Year	ID	Intercept	RE[a]	Residual	County	SSRE	SURE
2016	1	0.8109	−0.14853	0.10938	Aransas	−0.52056	0.38203
2016	2	0.8109	−0.93856	0.01142	Calhoun	−0.34658	−0.59198
2016	3	0.8109	−1.83254	0.00591	Goliad	−1.78520	−0.04734
2016	4	0.8109	1.73244	0.35783	Nueces	1.22239	0.51004
2016	5	0.8109	0.55809	0.09022	San Patricio	1.21657	−0.65849
2016	6	0.8109	0.62913	0.12533	Victoria	0.22338	0.40574
2013	1	0.8109	−0.14853	−0.01783	Aransas		
2013	2	0.8109	−0.93856	0.00306	Calhoun		
2013	3	0.8109	−1.83254	0.00245	Goliad		
2013	4	0.8109	1.73244	0.04573	Nueces		
2013	5	0.8109	0.55809	0.00813	San Patricio		
2013	6	0.8109	0.62913	0.01290	Victoria		
2010	1	0.8109	−0.14853	−0.09151	Aransas		
2010	2	0.8109	−0.93856	−0.01448	Calhoun		
2010	3	0.8109	−1.83254	−0.00841	Goliad		
2010	4	0.8109	1.73244	−0.40364	Nueces		
2010	5	0.8109	0.55809	−0.09828	San Patricio		
2010	6	0.8109	0.62913	−0.13817	Victoria		

Note: based on the two PSA eigenvectors

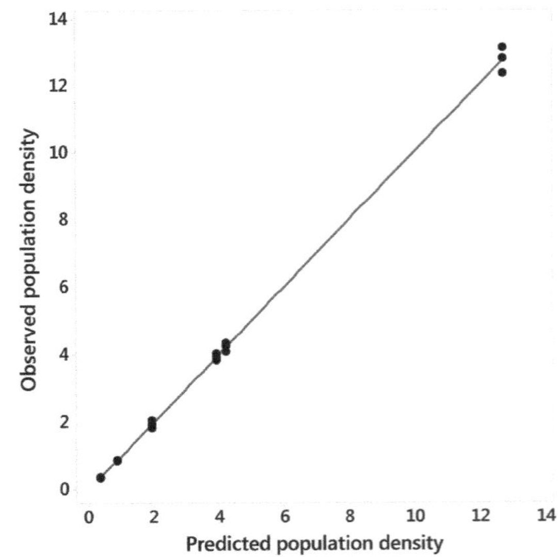

[a]SAS and R estimate slightly different random effects values (nevertheless, they align along a straight line in a scatterplot). Because estimation here involves nonlinear procedures, slight differences should be anticipated due to different algorithms, stopping rules, and machine precision.

Analysis of Variance					
Source	DF	Sum of Squares	Mean Square	F Value	Pr > F
Model	2	6.61276	3.30638	7.31	0.0703
Error	3	1.35701	0.45234		
Corrected Total	5	7.96977			

Root MSE	0.67256	R-Square	0.8297
Dependent Mean	0.00000658	Adj R-Sq	0.7162
Coeff Var	10223522		

Parameter Estimates						
Variable	Label	DF	Parameter Estimate	Standard Error	t Value	Pr > |t|
Intercept	Intercept	1	0.00000658	0.27457	0.00	1.0000
E_1		1	-1.60511	0.67256	-2.39	0.0970
E_5		1	2.00907	0.67256	2.99	0.0583

Fig. 8.13 Linear regression analysis results for the SSRE–SURE decomposition of the estimated RE term.

eigenvectors are significant. Nevertheless, for demonstrative purposes, an artificial SSRE component may be constructed by regressing the RE term on eigenvectors $\mathbf{E}_1$ and $\mathbf{E}_5$, the two extreme SA eigenvectors, respectively, portraying PSA and NSA; the resulting mixture SA linear combination accounts for roughly 83% of the variance in the RE term (hence the SURE component accounts for roughly 17% of the variance in the RE term). Fig. 8.13 is a screenshot of the linear regression output for this analysis; Table 8.9 employs its regression coefficients to illustrate the decomposition of RE into a SSRE and a SURE component.

The SSRE is a SA mixture, with the two eigenvectors portraying strong SA (Fig. 8.14): $\text{MC}=0.74$ and $\text{GR}=0.18$ for the PSA eigenvector, and $\text{MC}=-0.87$ and $\text{GR}=1.68$ for the NSA eigenvector.

These space–time data analyses cannot yet be executed with procedures in Spatial Analysis using ArcGIS Engine and R (SAAR) or ESF Tool. Appendix 8.A furnishes R code that allows replication of results for this toy example; similar code appears in Chapter 5 of Chun and Griffith (2013). The `functions.R` code is available at the URL appearing in Appendix 7.B. This code can be customized for a researcher's own spatial interaction data analysis. All calculation results appearing in this section were verified with SAS.

8.4 Summary

This chapter summarizes selected treatments of space–time data. It conceptualizes autocorrelation in space and time as a function of either some common factor comprising the sum of a SSRE and a SURE component that is constant across a time horizon, or a space–time structure term (an ESTF) coupled with a SURE component that is constant across a time horizon. Chapter 6 presents how to estimate a SSRE component; this chapter presents how to estimate a SURE component. This chapter includes a

Table 8.9 RE partitioning into SSRE and SURE components

ID	RE		a		E_1		E_5		SURE
1	−0.14853		0.0000		−0.33850		−0.53452		0.38112
2	−0.93856		0.0000		0.21592		0.00000		−0.59198
3	−1.83254	=	0.0000	−1.60511	0.44315	2.00738	−0.53452	+	−0.04825
4	1.73244		0.0000		−0.42523		0.26726		0.51340
5	0.55809		0.0000		−0.42523		0.26726		−0.66095
6	0.62912		0.0000		0.52988		0.53452		0.40665

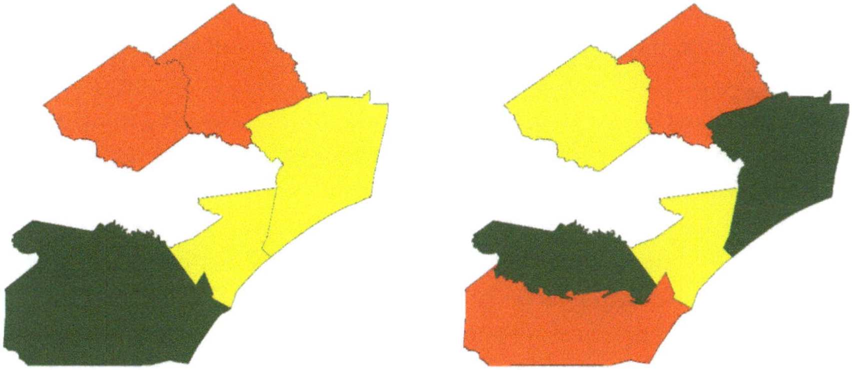

(A) (B)

Fig. 8.14 The RE components; the values of *dark green* to *yellow* to *dark red* are directly proportional to the eigenvector values. (A) PSA SSRE. (B) NSA SSRE.

demonstration that a RE = SSRE + SURE term can be robust to the set of time periods, as well as the frequentist or Bayesian approach, used to estimate it. This chapter also includes a demonstration that a constructed ESF tends to be reasonably stable across at least a short time period. In addition, this chapter extends ESF methodology to ESTF methodology, following two past MC_{ST} formulations, conceptualizing ESTFs in terms of lagged and contemporaneous space–time structures. These two conceptualizations can yield slightly different results. According to evidence appearing in this chapter, the contemporaneous specification appears to outperform the lagged and the common factor RE specifications; analyses of other georeferenced data elsewhere corroborate this finding (e.g., Griffith, 2013). The concluding substantive section of this chapter (Section 8.2.3) illustrates that including a SURE term with an ESTF term to calculate imputations noticeably reduces prediction error, even when half of a dataset is missing (at random). Meanwhile, the concluding section (Section 8.3) furnishes details of a toy analysis for replication purposes.

A departing aside for this chapter pertains to RE, this term's decomposition into SSRE and SURE components, and spatially varying coefficients. This chapter emphasizes the simplest case of an intercept being linked to a RE term. Chapter 6 (e.g., Section 6.2) addresses, in part, the case of spatially varying regression coefficients. Combining these two perspectives, a RE term could be added to each regression coefficient, too. MESF, in its simplest form, reveals that spatial structure in residuals, which is equivalent to the SSRE term in a population, can be identified without repeated measures (e.g., a space–time series) or Bayesian priors; residual SA furnishes the necessary additional information. Consequently, the resulting residuals are the sum of a SURE and a stochastic error term, ε; additional information is needed to separate these two components. Without additional information, estimation of a single-point-in-time SURE component becomes the residual term, which includes ε, and this stochastic error term becomes zero. Because estimating a parametric mixed—rather than a finite mixture of distributions—statistical model could stop with the estimation of a RE term, continuing by calculating its decomposition into a SSRE and a SURE component has several practical advantages. First, each of these two components represents missing variables in a model specification,[6] and hence each may furnish clues about which variables are missing. In a recent precision agricultural analysis (Long & Griffith, 2019), the RE term accounts for ~40% of the geographic variation in crop yield across a Georgia farm field (based on seven consecutive years of harvest data); 85% of this variation (i.e., 34% of the geographic variation in crop yield) relates to a SSRE component, and 15% (i.e., 6% of the geographic variation in crop yield) relates to a SURE component. The SSRE component has a map pattern that reflects the soil map for the field. The relatively weak SURE component is consistent with genetic variation in the random blend of seed stocks for a given crop variety. A second practical advantage of the RE decomposition concerns understanding impacts of SA. For example, to compare different geographic landscapes in terms of standard statistical assumptions (e.g., IID), the SSRE component needs to be isolated; this is analogous to estimating a different spatial autoregressive parameter, ρ, for each landscape. Similarly, simulating impacts of different natures and degrees of SA for a given geographic landscape also requires isolation of the SSRE component; this separation allows an exploration of the full range of SA impacts. In addition, separating an RE into its SSRE and SURE components enables a spatial scientist to assess impacts upon data analytic features (e.g., regression coefficient standard errors) attributable to SA. Therefore, readers should realize that this chapter merely introduces them to the mixed models framework, with special reference to parametric mixtures and SA, and that entire books (e.g., Lindsay, 1995) are devoted to this topic.

[6] Estimation could be concomitant with covariates, or a resulting RE term could be a response variable in a linear regression involving the covariates. Mixed statistical model theory argues for the former.

Appendix 8.A A Corpus Christi toy space–time dataset R code

```
## set a working directory
setwd("C:/IntroESF")

## load libraries and functions
library(rgdal)
library(spdep)
source("functions.R")

## read data
ccm <- read.csv("CorpusChristi_MSA-data.csv")
ccm <- ccm[order(ccm$id, ccm$year), ]
ccm$area <- ccm$area * 100000

cc.poly <- readOGR(dsn=getwd(), layer="CorpusChristi_MSA")
cc.nb <- poly2nb(cc.poly)

## run a GLMM
library(lme4)
library(optimx)
poi.mix <- glmer(population ~ 1 + (1|county), offset=log(area),
                 family = poisson, data = ccm, start=list(fixef=0),
                 control=glmerControl(optimizer="optimx",
                 optCtrl=list(method="nlminb")))
summary(poi.mix)
rande <- as.numeric(ranef(poi.mix)$county[,1]) # get random effects
rande

## generate eigenvectors
n <- length(cc.nb)
B <- nb2mat(cc.nb, style="B")
M <- diag(n) - matrix(1, n, n)/n
MBM <- M %*% B %*% M
eig <- eigen(MBM, symmetric=T)
EV <- as.data.frame(eig$vectors)
colnames(EV) <- paste("EV",1:n, sep="")
EV <- as.data.frame(EV[, round(eig$values, 4) != 0])

## run stepwise regression for ESF
lm.init <- lm(rande ~ 1, data=EV)
lm.full <- lm(rande ~ ., data=EV)
```

```
esf.ccm <- stepwise(lm.full, lm.init, sle=0.15, sls=0.1501,
verbose=T)

ssre <- fitted.values(esf.ccm)

sure <- residuals(esf.ccm)

## map the random effects

mapping.bipolar(cc.poly, ssre, 4, legend.loc="bottomright")

mapping.bipolar(cc.poly, sure, 4, legend.loc="bottomright")
```

References

Chun, Y., & Griffith, D. (2013). *Spatial statistics and geostatistics*. Thousand Oaks, CA: SAGE.

Cliff, A., & Ord, J. (1981). Spatial and temporal analysis: Autocorrelation in space and time. In N. Wrigley & R. Bennett (Eds.), *Quantitative geography: A British view* (pp. 104–110). London: Routledge & Kegan Paul.

Durbin, J., & Watson, G. (1951). Testing for serial correlation in least squares regression II. *Biometrika, 38*, 159–177.

Elhorst, J. (2010). Dynamic panels with endogenous interaction effects when T is small. *Regional Science and Urban Economics, 40*, 272–282.

Griffith, D. (1981). Interdependence in space and time: Numerical and interpretative considerations. In D. Griffith & R. MacKinnon (Eds.), *Dynamic spatial models* (pp. 258–287). Alphen aan den Rijn: Sijthoff and Noordhoff.

Griffith, D. (2008). A comparison of four model specifications for describing small heterogeneous space–time datasets: Sugar cane production in Puerto Rico, 1958/59–1973/74. *Papers in Regional Science, 87*, 341–356.

Griffith, D. (2012). Space, time, and space–time eigenvector filter specifications that account for autocorrelation. *Estadística Española, 54*(177), 7–34.

Griffith, D. (2013). Estimating missing data values for georeferenced Poisson counts. *Geographical Analysis, 45*, 259–284.

Griffith, D., & Heuvelink, G. (2012). Deriving space–time variograms from space–time autoregressive (STAR) model specifications. In A. Yeh, J. Shi, Y. Leung, & C. Zhou (Eds.), *Advances in spatial data handling and GIS* (pp. 3–14). Berlin: Springer-Verlag.

Lee, L.-F., & Yu, J. (2012). Spatial panels: Random components versus fixed effects. *International Economic Review, 53*, 1369–1412.

Lindsay, B. (1995). Mixture models: Theory, geometry, and applications. In *Vol. 5. NSF-CBMS Regional Conference Series in Probability and Statistics*, Hayward, CA: Institute of Mathematical Statistics.

Long, D., & Griffith, D. (2019). Interpreting spatial variation in multi-year yield data using eigenvector spatial filtering. *Paper Presented at the Annual Meeting of the American Association of Geographers*. Washington, DC, April 3–9.

Thomas, A., Best, N., Lunn, D., Arnold, R., & Spiegelhalter, D. (2004). *GeoBUGS user manual*. Version 1.2. Available at: https://www.mrc-bsu.cam.ac.uk/wp-content/uploads/geobugs12manual.pdf. Accessed 25 June 2018.

CHAPTER 9

MESF and multivariate statistical analysis

Chapter outline

This chapter extends content in the book by Griffith and Amrhein (1997), which illustrates the use of conventional multivariate statistics to analyze geospatial data but without accounting for spatial autocorrelation (SA). The goal of multivariate statistics is to address problems involving two or more variables, often a simultaneous testing situation, and accounting for the presence of multicollinearity (e.g., nonzero pairwise correlation) among a set of variables. This latter data feature relates multivariate analysis to SA in terms of correlated data. Primary techniques in multivariate statistics, other than multiple regression, include principal components analysis (PCA), factor analysis (FA), multivariate analysis of variance (MANOVA), discriminant function analysis (DFA), canonical correlation analysis (CCA), and cluster analysis (CA). Many treatments of multivariate statistical analysis exist; Johnson and Wichern (2007) furnish a particularly good overview.

Spatial Regression Analysis Using Eigenvector Spatial Filtering
https://doi.org/10.1016/B978-0-12-815043-6.00010-0

9.1 PCA, FA, and MESF

Redundant (duplicate) information contained in data—already introduced in terms of SA in Chapter 1—can enter through the inclusion of multiple attribute variables in an analysis, where the variables themselves are repetitious at least to some degree. Such redundant information causes multicollinearity to be present in covariate matrix **X**, and response variable–covariate pairs Y and X_j (j = 1, 2, …, P). One concern associated with the multivariate problem is avoiding double-counting of any repeated attribute information in covariates, and sometimes managing the replicate information common to response variable–covariate pairs. The techniques of PCA and FA address this covariates concern. The accompanying SA problem seeks to avoid double-counting any duplicated locational information, whose presence can alter (i.e., enhance or suppress) the degree of multicollinearity in a set of attribute variables. Moran eigenvector spatial filtering (MESF) addresses this spatial concern, paralleling, at least conceptually, PCA and FA (as mentioned in Section 3.1.1). Common SA map patterns tend to increase, whereas unique SA map patterns tend to decrease, the degree of multicollinearity detected in a set of attribute variables.

9.1.1 Selected mathematical features of PCA

The mathematical solution to the PCA problem involves eigenvalues and eigenvectors (Chapter 2) that are selected in a way to measure spread along a latent direction in the attribute data space by the variance of the areal unit attribute values in that direction. Given a set of P variables $\{X_j, j = 1, 2, \ldots, P\}$ converted to z-score mode (to remove the impact of measurement scale magnitude), and with $\mathrm{Det}(\mathbf{Z}^T\mathbf{Z}) \neq 0$, where Det denotes matrix determinant and superscript T denotes the matrix transpose operator, a vector of uncorrelated variables $\{\mathbf{F}_j, j = 1, 2, \ldots, P\}$ always can be derived by a set of linear transformations represented by matrix **E**, such that the diagonal matrix of λ_j eigenvalues is defined as follows:

$$\mathbf{\Lambda} = \mathbf{E}^T\left(\mathbf{Z}^T\mathbf{Z}\right)\mathbf{E}.$$

Matrix **E** is the eigenvector matrix extracted from the correlation matrix $\mathbf{R} = \mathbf{Z}^T\mathbf{Z} = \mathbf{E}\mathbf{\Lambda}\mathbf{E}^T$; the ordering of eigenfunctions in matrices **E** and **Λ** must be the same. Because the correlation matrix **R** is symmetric, the resulting eigenvectors are orthogonal, and the matrix **F** of synthetic variables $\{\mathbf{F}_j, j = 1, 2, \ldots, P\}$ is uncorrelated. In other words, the original attribute data space is rotated to a new one to maximize separation in this new synthetic data space; the objective is to find a more informative frame of reference within which to view a set of n areal unit attributes. Matrix **F** exhibits many appealing properties, with those of most interest in multivariate analysis relating to either unique or common variance; in contrast, MESF exhibits many appealing properties in terms of SA. Matrix **F** is what differentiates PCA from MESF: PCA employs eigenvectors to construct synthetic variates as linear combinations of original variables, X_j (i.e., the coefficients of a linear combination are its corresponding eigenvector elements), whereas MESF employs the eigenvectors themselves as synthetic variates. Griffith and Amrhein (1997, pp. 161–63) furnish set-theoretic and geometric graphical illustrations of these notions.

Principal component scores are the measures for the synthetic variates. They may be identified with the following algebra:

$$\mathbf{\Lambda} = \mathbf{E}^{\mathrm{T}}\mathbf{R}\mathbf{E} = \mathbf{E}^{\mathrm{T}}\left(\mathbf{Z}^{\mathrm{T}}\mathbf{Z}\right)\mathbf{E} = (\mathbf{Z}\mathbf{E})^{\mathrm{T}}(\mathbf{Z}\mathbf{E}),$$

where $\mathbf{E}^{\mathrm{T}}\mathbf{E} = \mathbf{I}$, and matrix $\mathbf{ZE}$ is the projection of the z-score matrix $\mathbf{Z}$ onto the new component axes. Matrix $\mathbf{ZE} = \mathbf{F}$ defines the mutually orthogonal linear combinations of the original variables. These are the component scores as well as the synthetic variates. Similarly,

$$\mathbf{R} = \mathbf{Z}^{\mathrm{T}}\mathbf{Z} = \mathbf{E}\mathbf{\Lambda}\mathbf{E}^{\mathrm{T}} = \left(\mathbf{E}\mathbf{\Lambda}^{1/2}\right)\left(\mathbf{E}\mathbf{\Lambda}^{1/2}\right)^{\mathrm{T}},$$

with $\mathbf{E}\mathbf{\Lambda}^{1/2}$ producing the component loadings or component structure matrix (the correlations between the original variables and the constructed synthetic variates). Typically this structure matrix identifies a first principal component reflecting the average magnitude of all P original variables, often accompanied by the remaining principal components basically representing contrasts between subsets of the original variables. Finally, because the sum of the eigenvalues equals the trace of matrix **R**, which is P (because all diagonal entries are one; that is, the correlation between X_j and itself), and this trace equals the total generalized variance embraced by the matrix **Z** of variables (each Z_j variable has a variance of one, by construction), λ_j/P ($j = 1, 2, \ldots, P$) may be interpreted as the percentage of variance accounted for by principal component j.

9.1.2 Multicollinearity

PCA addresses multicollinearity, or nonzero correlation among pairs and subsets of attribute variables. SA can impact upon this multicollinearity, inflating it when MESF eigenvectors are common, and deflating it when MESF eigenvectors are unique. Table 9.1 illustrates these impacts, with

Table 9.1 Selected sample correlation matrices (**Rs**) illustrating SA inflation/deflation impacts on multicollinearity based upon population density, percentage of unemployed, and log-income

Attribute variables	Aspatial components	Aspatial components + unique eigenvectors	Aspatial components + common eigenvectors
$\begin{pmatrix} 1 & -0.20 & 0.26 \\ -0.20 & 1 & -0.53 \\ 0.26 & -0.53 & 1 \end{pmatrix}$	$\begin{pmatrix} 1 & -0.13 & 0.02 \\ -0.13 & 1 & -0.37 \\ 0.02 & -0.37 & 1 \end{pmatrix}$	$\begin{pmatrix} 1 & -0.10 & 0.04 \\ -0.10 & 1 & -0.36 \\ 0.04 & -0.36 & 1 \end{pmatrix}$	$\begin{pmatrix} 1 & -0.27 & 0.28 \\ -0.27 & 1 & -0.47 \\ 0.28 & -0.47 & 1 \end{pmatrix}$
$\lambda_1/\lambda_3 = 3.58$	$\lambda_1/\lambda_3 = 2.26$	$\lambda_1/\lambda_3 = 2.16$	$\lambda_1/\lambda_3 = 3.18$

Note: The ratio λ_1/λ_n is the collinearity condition number, which quantifies the extent of multicollinearity in attribute data.

pairwise correlation coefficients appearing in its three-by-three correlation matrices, **R**: the ratio of the extreme eigenvalues (i.e., the collinearity condition number; see Montgomery, Peck, & Vining, 2001, p. 339) is greatest for the attribute variables (which contain SA), least for the combined aspatial components and unique eigenvector ESFs (deflation vis-à-vis the aspatial correlations), and inflated (vis-à-vis the aspatial correlations) with the inclusion of only common eigenvector ESFs.[1] Meanwhile, because SA inflates variance, and hence covariance, of georeferenced variables, and correlation coefficients are ratios of covariances and variances, limits on the range of the product moment correlation coefficient (i.e., ± 1) restrict this impact to some degree; in other words, some of the impacts in the numerator and the denominator divide out in the ratio. The basic question that MESF can address is whether SA alters loading patterns in a structure matrix, the signs of correlation coefficients appearing in a structure matrix, and/or the percentage of variance accounted for by a principal component.

In geographic landscapes lacking catastrophic disturbances, the SA latent in the geographic distribution of population density tends to be approximately the same from year to year for a relatively short time horizon. Consider the Texas counties for the seven-year period 2010–16. A PCA of these data should result in a single principal component that accounts for nearly all of the variation displayed by this set of attribute variables. Here the first principal component accounts for 99.98% of the spatial variation in log-population density. Table 9.2 discloses that the first principal component is the only one with prominent loadings, all seven being very close to one; a single dimension spans these geographic distributions. This aspatial analysis, however, fails to reveal the role of SA in these attribute variables.

Partitioning each attribute into its spatially structured map pattern, namely, its ESF term, and its spatially unstructured (i.e., aspatial) map pattern, namely, its residual term, allows an examination of the role of SA. Constructing ESFs with stepwise linear regression achieves this decomposition (see Chapter 3); the aspatial variates are the regression residuals. Table 9.3, paralleling Table 9.2, tabulates PCA summary results for both

[1] ESFs for these three attribute variables include four common eigenvectors accounting for an average of 14% of the spatial variation; the unique eigenvector ESFs account for an average of 13% of the spatial variation. The population density and log-income ESFs each contain seven unique eigenvectors; the percentage of unemployed ESF contains five unique eigenvectors. The correlation between the percentage of unemployed and log-income in this particular case is sufficiently close to zero that random noise tends to dominate deflation here (i.e., 0.02 and 0.04 mostly result from stochastic error, which tends to obscure any deflation).

Table 9.2 A PCA structure (i.e., loadings) matrix for annual Texas counties log-population density, 2010–16

Year	Principal component						
	I	**II**	**III**	**IV**	**V**	**VI**	**VII**
2010	**0.99984**	−0.01724	0.00315	0.00276	0.00300	−0.00110	−0.00034
2011	**0.99991**	−0.01267	0.00285	−0.00308	−0.00081	0.00274	0.00042
2012	**0.99995**	−0.00901	−0.00301	0.00151	−0.00360	−0.00141	0.00138
2013	**0.99999**	0.00113	−0.00188	−0.00423	−0.00014	−0.00174	−0.00200
2014	**0.99996**	0.00535	−0.00699	0.00254	0.00098	0.00188	−0.00086
2015	**0.99988**	0.01499	−0.00021	−0.00181	0.00235	−0.00055	0.00227
2016	**0.99982**	0.01745	0.00607	0.00232	−0.00179	0.00019	−0.00087

Note: A bold font denotes prominent component loadings (i.e., correlations; |r| >0.5).

Table 9.3 A PCA structure matrix for annual Texas counties log-population density ESFs and ESF residuals, 2010–16

Year	Principal component						
	I	II	III	IV	V	VI	VII
Annual ESFs[a]							
2010	**0.99986**	−0.01405	0.00926	0.00177	0.00109	−0.00050	−0.00039
2011	**0.99988**	−0.01344	0.00708	0.00018	−0.00081	0.00056	0.00093
2012	**0.99990**	−0.01331	0.00301	−0.00185	−0.00047	0.00084	−0.00076
2013	**0.99992**	−0.01209	−0.00235	−0.00112	−0.00095	−0.00147	0.00004
2014	**0.99989**	−0.01292	−0.00699	−0.00110	0.00189	0.00016	0.00038
2015	**0.99988**	−0.01080	−0.01044	0.00215	−0.00077	0.00040	−0.00020
2016	**0.99705**	0.07682	0.00044	−0.00002	0.00003	0.00000	0.00000
Annual ESF residuals[a]							
2010	**0.99956**	−0.02130	0.01931	0.00082	0.00636	0.00061	−0.00158
2011	**0.99970**	−0.01950	0.01250	0.00543	−0.00314	−0.00318	0.00249
2012	**0.99977**	−0.01857	0.00762	−0.00578	−0.00318	0.00390	0.00188
2013	**0.99985**	−0.01446	−0.00662	0.00300	−0.00409	0.00063	−0.00444
2014	**0.99980**	−0.01333	−0.01187	−0.00820	0.00107	−0.00359	−0.00030
2015	**0.99965**	−0.00876	−0.02434	0.00481	0.00302	0.00165	0.00197
2016	**0.99534**	0.09634	0.00342	−0.00008	−0.00003	−0.00001	−0.00001

[a] Variable $= b_0 + ESF_{variable} + e$, where e denotes the attribute variable ESF residuals; the aspatial component is $b_0 + e$.
Note: A bold font denotes prominent component loadings (i.e., correlations; $|r| > 0.5$).

the ESFs and their corresponding aspatial residuals. The respective percentages of variance accounted for in the sets of ESFs and residuals by their respective principal component are 99.90% and 99.82%. Both cases have a single dimension spanning their corresponding sets of variates, very similar to their parent original georeferenced data (Table 9.2).

In contrast, rather distinct attribute variables may well contain different SA map patterns. Table 9.4 summarizes the principal components structure for the following set of attribute variables described in Table P1 (here density is based upon physical land area rather than total polygon area): population density, female–male ratio, median income, and percentages of white, African American, and Hispanic, commuters by car, commuters by public transport, high school graduates, unemployed, owned homes, and vacant homes. For simplicity, while better aligning each marginal frequency distribution with a bell-shaped curve, except for female–male ratio and percentages of white, high school graduates, unemployed, and owned homes, each variable was subjected to a Box–Cox logarithmic transformation; because of the presence of zeros, adding 0.6 to the percentage of African Americans and 0.002 to the percentage of public transit commuters was done prior to applying this transformation. These data have at most six dimensions (because principal component VII has no prominent loadings), with the first seven principal components accounting for a total of 91.93% of the spatial variation (the first six account for 86.84%); Table 9.4 lists the respective sequential component variance accounted for percentages. Table 9.4 also summarizes the accompanying component structure matrix for this analysis; this table does not have the type of simple structure appearing in Table 9.2. Because the seventh component has no prominent loadings, it can be classified as not being a dimension, reducing the number of dimensions from seven to six (accounting for a total of 86.84% of the spatial variation).

Again, partitioning each attribute into its spatially structured ESF term and its spatially unstructured linear regression residuals allows an examination of the role of SA. The map pattern components reduce to four, rather than represent the original six dimensions, whereas the aspatial components expand to seven dimensions. The respective percentages of variance accounted for are 89.09% by four components in the former case, and 86.61% by seven components in the latter case. The principal components structure (i.e., loadings or correlations) of both the ESF map patterns and the spatially unstructured map patterns (Table 9.5) is rather dissimilar to that appearing in Table 9.4 with regard to prominent loadings and number of mathematical dimensions. The combination of these two factors shifts

Table 9.4 A PCA structure matrix for selected 2010 Texas counties socioeconomic/demographic attribute variables

	Principal component						
Attribute	**I**	**II**	**III**	**IV**	**V**	**VI**	**VII**
Population density	**0.80805**	0.00816	−0.29893	0.18739	0.18541	−0.07964	0.09586
Female/male	0.31921	0.10706	0.38714	**0.73883**	0.11299	0.25139	−0.29045
% White	−0.45478	0.16528	**0.65227**	0.19666	0.33429	−0.13812	0.30533
% African American	**0.51685**	−0.48826	**−0.56696**	0.02019	−0.06503	0.26213	−0.02801
% Hispanic	−0.31842	**0.74256**	−0.39697	−0.11546	0.08650	−0.16086	−0.22531
% Car commuters	−0.47116	0.42898	−0.04798	−0.02469	−0.38757	**0.54749**	0.35283
% Public transit commuters	0.41075	0.28938	−0.39618	0.46189	−0.18111	−0.35071	0.39961
% High school graduates	**0.55284**	−0.47035	0.49980	−0.00237	−0.35219	−0.02723	0.12410
% Unemployed	−0.47987	**−0.61693**	−0.29677	−0.09471	0.41477	−0.01254	0.21995
% Owned homes	**0.59603**	0.19463	0.18760	−0.25878	**0.57771**	0.24295	0.13691
% Vacant homes	**−0.78150**	−0.45990	0.08920	0.16399	−0.14459	−0.15700	−0.07666
Median income	**0.62751**	0.15144	0.41520	−0.49103	−0.23029	−0.19421	−0.00896
% Variance accounted	30.14	16.42	15.47	9.93	8.81	6.07	5.09

Note: A bold font denotes prominent component loadings (i.e., correlations; $|r| > 0.5$).

Table 9.5 A PCA structure matrix for selected 2010 Texas counties socioeconomic/demographic attribute variable ESFs and ESF residuals

Attribute	Principal component						
	I	II	III	IV	V	VI	VII
Attribute ESFs[a]							
Population density	**0.83473**	0.15575	−0.13917	0.23397	−0.26561	−0.02780	−0.09048
Female/male	0.39499	0.01954	0.38622	**0.66137**	0.26436	−0.32115	−0.23931
% White	**−0.59570**	−0.27929	0.48264	0.41881	−0.25356	0.14526	0.15592
% African American	**0.65730**	**0.55478**	−0.23659	−0.26149	0.21038	−0.15595	−0.01709
% Hispanic	−0.44957	**−0.53859**	−0.58821	0.18863	−0.02134	0.06255	−0.18813
% Car commuters	**−0.66660**	−0.29388	0.12552	−0.22940	0.48338	−0.21398	0.21078
% Public transit commuters	0.41376	0.23952	−0.42685	**0.51161**	0.29716	0.36368	0.31392
% High school graduates	**0.62219**	0.22642	**0.62899**	−0.15185	0.05871	0.24923	0.02546
% Unemployed	**−0.52867**	**0.65795**	−0.09397	−0.08464	−0.33698	−0.14419	0.11041
% Owned homes	**0.71636**	−0.39507	0.03243	0.02390	−0.19065	−0.30122	0.39447
% Vacant homes	**−0.73895**	**0.52202**	0.20411	0.05333	0.10262	0.16105	−0.03382
Median income	**0.63978**	**−0.51609**	0.13916	−0.37974	0.02404	0.27758	−0.10137
% Variance accounted	38.27	16.86	12.23	10.53	6.13	5.07	3.74
Attribute ESF residuals[a]							
Population density	**0.53079**	**0.58287**	0.18560	0.06739	0.31216	−0.00368	0.09559
Female/male	−0.28548	0.41964	−0.43785	−0.14124	**0.53425**	0.07876	−0.42545
% White	**−0.60167**	0.08217	−0.39335	0.36425	0.11937	0.32057	0.28392
% African American	**0.61387**	−0.08306	**0.56092**	0.01497	0.29179	0.13302	−0.15441
% Hispanic	**0.68704**	0.01947	−0.38661	0.16905	−0.32555	−0.22318	−0.12185
% Car commuters	0.44150	0.17332	−0.15666	−0.28910	−0.36027	**0.72331**	−0.00029

% Public transit commuters	0.38009	0.34348	−0.27121	−0.36594	0.24196	−0.15715	**0.62232**
% High school graduates	**−0.52939**	0.36453	**0.51876**	−0.36689	0.08021	0.11827	0.04469
% Unemployed	0.15745	**−0.67970**	0.29626	0.43088	0.23468	0.18986	0.21665
% Owned homes	−0.06184	0.48748	0.00550	**0.76713**	0.00444	0.05621	0.02405
% Vacant homes	−0.32917	**−0.80579**	−0.11756	−0.25096	0.06980	−0.01563	0.06946
Median income	−0.46874	**0.52205**	0.37765	0.02728	−0.42465	−0.12892	0.09682
% Variance accounted	21.25	20.27	12.14	1.47	8.53	6.62	6.34

[a] Variable $= b_0 + ESF_{variable} + e$, where e denotes the attribute variable ESF residuals; the aspatial component is $b_0 + e$.
Note: A bold font denotes prominent component loadings (i.e., correlations; $|r| > 0.5$).

the percentage of white population from the principal component. They also mask the prominently common map patterns of other principal component variables with those for the percentages of car commuters and unemployed. Meanwhile, the aspatial principal component better captures ethnicity, separating it from income and housing variables. In addition, these two sets of variates also have some parts of their respective multicollinearity structures that are conspicuously different. The original PCA has only two variables that correlate strongly with more than one principal component. The ESF PCA has five such pairings, whereas the aspatial residuals PCA has three.

9.1.3 Moving from PCA to FA: Seeking parsimony

FA can be viewed as a technique through which a few (i.e., $Q << P$) prespecified synthetic variables are constructed so that, once all of these hypothetical variables are controlled for, pairwise correlations between the residual parts of the original P variables become zero. One fundamental difference between PCA and FA, then, is that PCA is concerned with accounting for variances of the original P variables solely through orthogonal axis rotation, whereas FA is concerned with explaining correlations among these original variables with a model consisting of a posited number of common factors. Another basic difference is that PCA extracts mathematical dimensions sequentially, maximizing component variance at each step, whereas FA extracts mathematical dimensions in a way that simultaneously (i.e., jointly) maximizes dimensional variances. Many facets of FA exist. Only one will be discussed here. The initial conceptualization of PCA more than 100 years ago by Pearson (1901) anticipated that it would produce results like those appearing in Table 9.2: multicollinearity in a covariate matrix **X** would render a single principal component that summarized almost all of its variance. But practice proved otherwise; most real-world data results are like those appearing in Table 9.4. The former is simple structure (e.g., a variable correlates highly with one and only one principal component; Revelle & Rocklin, 1979, p. 405) by definition, whereas the latter often defies being described as simple structure. This outcome advocates for a conversion from PCA to FA.

One form of FA begins with a PCA of the data matrix **X**. PCA constructs components in a sequential way such that the first component accounts for a maximum of common variance, the second component accounts for a maximum of residual common variance (once the original variance has been adjusted so that the variance accounted for by the first component is removed) subject to its being orthogonal to the first, and so on. But the resulting solution is not necessarily interpretable. Moreover, this criterion of sequential variance maximization is somewhat arbitrary, identifying only the principal patterns of variation without necessarily identifying distinctive clusters of variables. Rotating a principal components solution often addresses these weaknesses by enhancing interpretability and uncovering distinctive clusters of variables. In terms of interpretability, such simplicity is a preferred solution. This rotation takes an analysis from PCA into the realm of FA. It also highlights another difference between PCA and MESF: the sequential decomposition of a spatial weights matrix for MESF needs to be preserved, for example, to maintain the possible extreme degrees of SA—SA eigenfunctions are never revised with simple structure rotations.

Rotation replaces the sequential variance maximization criterion with some alternate objective criterion that attempts to simplify the revealed components structure (i.e., loadings) matrix in some unique manner. Once rotated, this matrix becomes the rotated factor structure matrix. Various principles of simple structure can be invoked to achieve this end. Generally speaking, these principles should attempt to alter the initial loadings matrix in such a way that: (1) each row has at least one of its loadings (i.e., correlation coefficients) close to one in absolute value, (2) at least as many variables as the number of factors minus one have near-zero loadings for each row, (3) every pair of columns has several variables with high loadings in one but not in the other, (4) a large proportion of the variables has near-zero loadings in most entries of the structure matrix, and (5) only a small number (preferable none) of column pairs have high nonzero loadings for the same variables. VARIMAX [compared with QUARTIMAX, EQUAMAX (which seldom is spelled EQUIMAX), and ORTHOMAX, for example] rotation is a widely used and appealing principle in multivariate geospatial applications. Not only does it select the set of eigenvectors so that all variances accounted for are maximized simultaneously, but its rotation also emphasizes column simplification in the structure matrix.

Algebraically, the rotation problem is one of finding some matrix $\mathbf{L}$, such that $\mathbf{L}^T\mathbf{L}=\mathbf{I}$ (the rotation matrix is orthogonal and its columns are normalized). This matrix then is applied to the PCA solution such that

$$\mathbf{L}^T\boldsymbol{\Lambda}\mathbf{L}=\mathbf{L}^T\mathbf{E}^T\mathbf{R}\mathbf{E}\mathbf{L}=\mathbf{L}^T\mathbf{E}^T\left(\mathbf{Z}^T\mathbf{Z}\right)\mathbf{E}\mathbf{L}=(\mathbf{E}\mathbf{L})^T\left(\mathbf{Z}^T\mathbf{Z}\right)(\mathbf{E}\mathbf{L}).$$

Because matrix $\mathbf{L}$ operates on the diagonal matrix of eigenvalues, $\boldsymbol{\Lambda}$, it will cause a redistribution of the value P over the diagonal entries. Because PCA sequentially extracts the maximum eigenvalues possible, subject to their summing to P, then matrix **L** tends to decrease the larger, and increase the smaller, eigenvalues. These new eigenvalues have the corresponding eigenvectors **EL** associated with them. Literally an infinite number of possible **L** matrices exist, although the four aforementioned structure (rotation) principles allow most possibilities to be dismissed.

The preceding Texas annual population density data yield a single component and hence do not need to be subjected to a VARIMAX rotation to achieve simple structure. In contrast, the 2010 Texas socioeconomic/demographic data fail to display simple structure in their initial solution (e.g., two variables have prominent loadings on two components, with none exceeding ± 0.9) and may well benefit from a VARIMAX rotation. Preliminary results appearing in Table 9.4 suggest that only six components need to be rotated; Table 9.6 presents a simpler structure, with variance accounted for by the principal factor decreasing from 30.14% to 18.03%, and that by the sixth factor increasing from 6.07% to 10.34%, one variable has a correlation exceeding 0.9, with only one variable having prominent loadings on two factors.

Meanwhile, a VARIMAX rotation seems to more effectively articulate SA map pattern alignments in the socioeconomic/demographic data for the spatially structured (i.e., ESFs) loadings matrix than aspatial information in the spatially unstructured (i.e., residual random map pattern) loadings matrix (Table 9.7). For the former case, five variables load onto two principal components before rotation, whereas no factor dimensions have multiple loadings after rotation (i.e., the one and only one factor loading principle). For the latter case, three variables load onto two principal components before rotation, with one still loading onto two factor dimensions after rotation.

9.2 MANOVA and MESF

MANOVA extends univariate analysis of variance (ANOVA) to two or more variables, with a goal of accounting for the accompanying multicollinearity and multiple testing complications. In its simplest form, it generalizes a two-sample difference of regional means test to K samples of multiple variable vectors of regional means, each with Q means rather than one mean (the problem addressed by ANOVA), with the accompanying assumptions of homogeneity within regions' covariance matrices and within regional means multivariate normal frequency distributions. This K-samples problem can be solved by writing the MANOVA model as a general linear regression

Table 9.6 A VARIMAX-rotated FA structure matrix for selected 2010 Texas counties socioeconomic/demographic attribute variables

	Rotated factor					
Attribute	**I**	**II**	**III**	**IV**	**V**	**VI**
Population density	**0.55131**	−0.00435	−0.20952	0.38986	0.30283	−0.47606
Female/male	−0.12804	−0.21546	−0.10566	0.14407	**0.88896**	−0.00967
% White	**−0.87851**	0.00006	0.13430	−0.00654	0.19796	−0.00795
% African American	**0.88042**	−0.24023	0.20019	0.11605	0.03487	−0.11280
% Hispanic	−0.10026	**0.88735**	−0.16109	−0.00629	−0.14452	0.11293
% Car commuters	−0.06188	0.28832	−0.08884	−0.17974	0.03546	**0.85468**
% Public transit commuters	0.39661	0.34525	−0.39000	−0.19059	0.39250	−0.39948
% High school graduates	0.09861	**−0.86302**	−0.33908	0.00396	0.06739	−0.16551
% Unemployed	0.02085	−0.02537	**0.87263**	−0.13271	−0.28481	−0.13715
% Owned homes	0.00476	−0.07018	−0.09065	**0.92315**	0.06804	−0.13438
% Vacant homes	−0.33770	−0.13427	**0.50971**	**−0.69721**	−0.13030	0.08785
Median income	−0.02696	−0.36971	**−0.72782**	0.36350	−0.30859	−0.16266
% Variance accounted	18.03	16.66	16.44	14.52	10.84	10.34

Note: Although the unrotated SAS and R solutions are identical, their Varimax-rotated solutions differ, because of such factors as algorithm differences (e.g., stopping rules and precision). The sign pattern for a given factor can be reversed because eigenvectors are unique except for a multiplicative constant of −1. A bold font denotes prominent component loadings (i.e., correlations; $|r| > 0.5$).

Table 9.7 A VARIMAX-rotated structure matrix for annual Texas counties log-population density ESFs and ESF residuals, 2010–16

	Rotated factor						
Attribute	**I**	**II**	**III**	**IV**	**V**	**VI**	**VII**
Attribute ESFs[a]							
Population density	−0.38402	**0.68917**	0.27032	−0.31565			
Female/male	−0.17290	**0.60503**	0.36049	0.46592			
% White	0.20415	−0.20126	−0.06915	**0.86847**			
% African American	−0.02662	0.36675	0.37813	**−0.76549**			
% Hispanic	−0.06902	−0.08643	**−0.91264**	0.16918			
% Car commuters	0.17286	**−0.63670**	−0.27044	0.30116			
% Public transit commuters	−0.00339	**0.78725**	−0.14325	−0.17950			
% High school graduates	−0.27463	0.09936	**0.87604**	−0.06593			
% Unemployed	**0.82972**	−0.11495	0.00261	−0.16348			
% Owned homes	**−0.75702**	0.25950	0.15271	−0.08452			
% Vacant homes	**0.86699**	−0.23928	0.06318	0.22391			
Median income	**−0.85399**	−0.15091	0.22146	−0.19524			
% Variance accounted	25.60	18.49	17.68	16.12			
Attribute ESF residuals[a]							
Population density	**0.60866**	0.00176	−0.39901	0.05533	0.45112	0.14860	0.09344
Female/male	0.09043	0.09778	0.16242	−0.19281	0.03291	**0.92606**	−0.04397
% White	0.12583	0.16567	**0.87242**	0.14196	−0.03765	0.16639	−0.01158
% African American	0.19666	0.03174	**−0.74081**	0.46845	0.02312	−0.03375	0.12518
% Hispanic	0.16045	**−0.82640**	−0.20301	−0.16970	0.08942	−0.12144	0.14024
% Car commuters	0.03376	−0.11599	−0.07258	−0.05667	0.08650	−0.03240	**0.97696**

% Public transit commuters	0.03239	−0.08776	−0.01622	−0.09770	**0.95688**	0.02106	0.07628
% High school graduates	0.02588	**0.88113**	−0.05854	−0.23915	−0.00224	0.03966	−0.00421
% Unemployed	−0.09752	−0.07375	−0.01566	**0.86361**	−0.16004	−0.31062	−0.12306
% Owned homes	**0.83285**	−0.05730	0.30892	0.05801	−0.14924	−0.00139	−0.12548
% Vacant homes	**−0.82049**	0.02260	0.19769	0.28072	−0.14907	−0.06370	−0.16284
Median income	0.34336	**0.52252**	0.13446	**−0.54449**	−0.14163	−0.32727	−0.09646
% Variance accounted	16.30	15.01	14.15	12.53	10.23	9.45	8.94

[a]Variable $= b_0 + ESF_{variable} + e$, where e denotes the attribute variable ESF residuals; the aspatial component is $b_0 + e$.

Note: Although the unrotated SAS and R solutions are identical, their Varimax-rotated solutions differ, because of such factors as algorithm differences (e.g., stopping rules and precision). The sign pattern for a given factor can be reversed because eigenvectors are unique except for a multiplicative constant of −1. A bold font denotes prominent component loadings (i.e., correlations; $|r| > 0.5$).

model whose specification includes binary indicator variables representing group (i.e., regional) membership for each areal unit, with any one of these indicator variables being chosen arbitrarily and either deleted from the analysis or subtracted from the remaining $K-1$ indicator variables. In addition, an n-by-1 vector of ones, denoted **1**, is included as the variable associated with the intercept term. In other words, each row sum of the set of K indicator variables equals one, which appears in the vector **1** for the intercept term, resulting in perfect multicollinearity, requiring one of the K indicator variables to be set aside (i.e., have its regression coefficient set to zero).

The response variable **Y** is an n-by-Q matrix, where Q denotes the number of response variables. The **X** matrix consists of the vector **1** and the $K-1$ indicator variables (i.e., $P=K$). MESF eigenvectors can be added to this **X** matrix to account for SA, filtering it out of the regression residuals and transferring it to the mean response, which becomes spatially varying. MANOVA then can assess differences of regional means in terms of either SA map pattern variates (i.e., ESFs) or aspatial (i.e., SA map pattern free) variates. Consider the 77 Texas counties classified as metropolitan statistical areas (MSAs) by the United States (US) Census Bureau (see Fig. 5.1A). The null hypothesis for the accompanying MANOVA problem may be written as

$$H_0: \boldsymbol{\mu}_{MSA} = \boldsymbol{\mu}_{not-MSA},$$

where $\boldsymbol{\mu}_j$ is a 12-by-1 vector of attribute variable population regional means. The constant variance assumption here is that the 12-by-12 within-regions covariance matrices for the two regional groups are equal; a diagnostic test assessment of this hypothesis rejects it, implying that the within-regions covariance matrices need to be pooled in order to test the equality of the multivariate regional means hypothesis. Individual attributable variable tests—for which separate ANOVAs can yield any one of the 2^Q possible combinations of univariate null hypothesis decisions (i.e., reject or fail to reject H_0) for Q variables—reveal that only the average percentage of Hispanics is not significantly different between the two regions; this finding holds with an adjustment for nonconstant variance [i.e., a Welch's (1951) test], which is necessary only for the female–male ratio. Among other data features, these individual null hypothesis tests posit an assumption about the diagonal, but not about the off-diagonal, entries of the covariance matrices. The MANOVA results indicate a highly significant difference between the multivariate means, with Wilk's lambda$=0.49$, Pillai's trace$=0.51$, the Hotelling–Lawley trace$=1.04$, and Roy's greatest/largest root$=1.04$

(see, e.g., Johnson & Wichern, 2007, p. 398). After adjusting for SA, these values become the following: Wilk's lambda = 0.73, Pillai's trace = 0.27, the Hotelling–Lawley trace = 0.36, and Roy's greatest root = 0.36. This same adjustment reduces the homogeneity of within-regions covariance matrices chi-square statistic from 329.4 to 248.5. Nevertheless, the vectors of regional means remain significantly different, with more inconsistencies appearing between this joint result and the 12 individual ANOVA results (see Table 9.8).

9.3 DFA and MESF

In various ways, DFA is an extension of MANOVA, with a focus on classification of areal units rather than on whether their multivariate means differ. Furthermore, a two-regions (i.e., two-groups) case can be conceptualized as the following linear regression problem:

$$\mathbf{Y} = \mathbf{X}\mathbf{b} + \mathbf{e} = \sum_{j=1}^{P} \mathbf{E}_{k,j}\mathbf{b}_{k,j} + \mathbf{X}^{*}\mathbf{b} + \mathbf{e},$$

where $\mathbf{Y}$ is a binary 0-1 response variable (with a vector dimension of n-by-1) indicating membership in one of the two regions, $\mathbf{E}_{k,j}\mathbf{b}_{k,j}$ is the jth ESF, and $\mathbf{X}^{*}$ is the aspatial part of matrix $\mathbf{X}$; here an ESF_j is constructed for each $\mathbf{X}_j$ covariate, separately, resulting in P ESFs. Spatial scientists repeatedly encounter or are called upon to regionalize areal data. DFA can be an appropriate analytical technique for such undertakings when several distinct parent populations (e.g., regions) of areal units are known and expected to exist. This situation is in stark contrast to one in which prespecified multiple populations are suspected to exist but are unknown (i.e., the problem addressed by CA). In this latter case, the classification chore becomes one of uncovering latent natural classes (or regions) by sorting areal units into potential regional populations.

The DFA classification task addresses two distinct goals. First, a given population classification itself can be evaluated. The initial step in such classification evaluation is MANOVA for the multivariate case (the topic of the preceding section). If MANOVA indicates the presence of significant differences between multivariate regional means, then the relevant question is which variables contribute to the detected average regional differentiation. The curiosity motivating this question is similar to that prompting the MANOVA question of which of the P individual regional means lead to rejection of its null hypothesis. Because of the inclusion of P variables here,

Table 9.8 H_0 hypothesis testing probabilities for individual univariate results and for their accompanying joint multivariate results

	Georeferenced data		ESFs		Aspatial residuals	
Univariate attribute variable	**ANOVA**	**Levene**	**ANOVA**	**Levene**	**ANOVA**	**Levene**
Population density	**<0.0001**	0.1419	**<0.0001**	0.8105	**<0.0001**	0.0536
Female/male	**0.0004**	**0.0044**	**0.0012**	**0.0002**	0.0376	0.0288
% White	**0.0011**	0.5272	**<0.0001**	0.6605	0.8055	0.3010
% African American	**0.0004**	0.0648	**<0.0001**	0.5871	0.5673	0.4421
% Hispanic	0.2044	0.1509	0.3027	0.0618	0.4542	0.2399
% Car commuters	**<0.0001**	0.1956	**<0.0001**	0.6832	0.0501	0.8666
% Public transit commuters	**<0.0001**	0.4613	**0.0035**	0.7301	**<0.0001**	0.7395
% High school graduates	**<0.0001**	0.1061	**<0.0001**	0.5377	**0.0004**	0.2163
% Unemployed	**<0.0001**	0.1368	**<0.0001**	0.8889	**0.0002**	0.0315
% Owned homes	**<0.0001**	0.9431	**<0.0001**	0.3672	0.0536	0.8064
% Vacant homes	**<0.0001**	0.5622	**<0.0001**	0.0883	**<0.0001**	0.3632
Median income	**<0.0001**	0.5328	**<0.0001**	0.1886	**<0.0001**	0.2805
Multivariate variables	MANOVA	$\mathbf{W}_{MSA} = \mathbf{W}_{not\text{-}MSA}$	MANOVA	$\mathbf{W}_{MSA} = \mathbf{W}_{not\text{-}MSA}$	MANOVA	$\mathbf{W}_{MSA} = \mathbf{W}_{not\text{-}MSA}$
12 variables jointly	**<0.0001**	**<0.0001**	**<0.0001**	**<0.0001**	**<0.0001**	**<0.0001**

Note: A bold font denotes statistically significant results at the 0.01 level.
Matrix $\mathbf{W}_j$ denotes the within regions covariance matrix for region j.

this scenario parallels the multiple comparisons notion affiliated with the testing of separate univariate equality of regional means hypotheses. The relationship between individual ANOVAs and MANOVA is further complicated by the presence of multicollinearity among the P variables. DFA particularly addresses these two features of MANOVA. As such, it furnishes supplemental data analysis results involving linear combinations of attribute variables that maximize the differences of means between regions. In addition, the importance of attribute variables can be assessed with a stepwise selection procedure, albeit one compromised by multicollinearity.

Table 9.9 summarizes stepwise results for the MANOVA data analysis described in the preceding section. Population density and median income are common variables across the three analyses. Adjusting for SA makes a difference: SA is the source of most of the regional differentiation attributable to population density and median income; spatially unstructured differentiation requires more attribute variables; and SA contributes roughly half or more of the differentiation power in this example.

DFA determines optimal linear combinations of variables, much like those computed with PCA or FA, that reveal the dimensions along which the most conspicuous regional mean differences occur. Just as in PCA or FA, a structure matrix can be constructed for DFA revealing the nature and degree of relations between the original variables and these discriminating dimensions, suggesting the directions along which these most conspicuous regional mean differences occur. Table 9.9 reports those variables with significant DFA structure matrix loadings for the example analyzed here.

9.3.1 The DFA eigenfunction problem

The presence of multiple variables and their covariances results in the generalization of ANOVA between-regions and within-regions variances to between-regions and within-regions covariance matrices, respectively, denoted by **B** and **W**; the respective individual ANOVA variances are the diagonal entries of these matrices. Because matrices are involved, DFA has the following associated eigenfunction problem:

$$\text{eigenvalues}: \text{Det}\left(\mathbf{W}^{-1}\mathbf{B} - \lambda\mathbf{I}\right) = 0, \text{and} \tag{9.1}$$

$$\text{eigenvectors}: \left(\mathbf{W}^{-1}\mathbf{B} - \lambda_j\mathbf{I}\right)\mathbf{E}_j = \mathbf{0},\ \mathbf{E}_j \neq \mathbf{0}. \tag{9.2}$$

Testing for the difference of multivariate regional means is based upon the solution to Eq. (9.1), whereas DFA scores employ the eigenvectors $\mathbf{E}_j$ that are the solutions to Eq. (9.2).

Table 9.9 Significant stepwise DFA variable selection, $\alpha = 0.10$

Original attribute variables			Attribute variable-specific ESFs (spatially structured)			Attribute variable-specific ESF residuals (spatially unstructured)		
Attribute	**Partial R^2**	**Wilk's lambda**	**ESF**	**Partial R^2**	**Wilk's lambda**	**Residuals**	**Partial R^2**	**Wilk's lambda**
Population density	0.3853	0.6147	Population density	0.2706	0.7294	Population density	0.1185	0.8815
Median income	0.1360	0.5311	Median income	0.0570	0.6878	Median income	0.0657	0.8236
Female–male ratio	0.0177	0.5216	% Hispanic	0.0134	0.6786	% Public commuters	0.0466	0.7852
			% High school graduates	0.0119	0.6706	% Car transit commuters	0.0421	0.7522
			% Vacant houses	0.0139	0.6613	% High school graduates	0.0139	0.7418

Note: The Wilk's lambda H_0 probability is <0.0001 for all variables except the female–male ratio, for which it is 0.0346.

9.3.2 DFA as a regression problem: Two-regions DFA

Inspecting the two-regions (K=2) DFA case illustrates the relationship between linear DFA and linear regression. Advantages to studying this relationship are that the more demanding problem of computing the eigenfunctions of matrix $\mathbf{W}^{-1}\mathbf{B}$ can be avoided, and that a direct connection with SA can be established. In principle, the simplification that materializes arises, in part, from the existence of a single nonzero eigenfunction for matrix $\mathbf{W}^{-1}\mathbf{B}$; moreover, MIN(K−1, P)=1 for K=2. Tatsuoka and Lohnes (1988, pp. 225–231) outline the algebraic manipulations necessary to move from $(\mathbf{W}^{-1}\mathbf{B}-\lambda_1\mathbf{I})\mathbf{E}_1=\mathbf{0}$ to $(\mathbf{X}^T\mathbf{X})^{-1}\mathbf{X}^T\mathbf{Y}$, where matrix $\mathbf{X}$ in the regression model contains the set of covariates used to construct the DFA classifier (e.g., the stepwise selected variables), and vector $\mathbf{Y}$ is the binary (0-1) classification variable used to partition the DFA covariance matrices that sum to $\mathbf{W}$ and $\mathbf{B}$. Two noteworthy properties are: (1) $\mathbf{1}^T(\mathbf{X}-\mathbf{1}\overline{\mathbf{X}}^T)=\mathbf{0}$ (which has vector dimension 1-by-P) implies that the matrix used to construct matrices $\mathbf{W}$ and $\mathbf{B}$ is orthogonal to the vector $\mathbf{1}$ used in regression to compute an intercept term; and (2) $\mathbf{E}_1$ of $\mathbf{W}^{-1}\mathbf{B}$ is proportional to the regression coefficient vector $\mathbf{b}=(\mathbf{X}^T\mathbf{X})^{-1}\mathbf{X}^T\mathbf{Y}$.

Table 9.10 tabulates the eigenvector elements for the single nonzero eigenvalue of the DFA matrix $\mathbf{W}^{-1}\mathbf{B}$ as well as the linear regression coefficients for those attribute variables and synthetic variates appearing in Table 9.9. For each DFA, the ratio of the pairs of values yields a constant (except for minor rounding error) for each analysis. One primary outcome of this relationship is that SA can be accounted for in a simple DFA by a MESF-constructed ESF.

9.4 CCA and MESF

Traditional regression analysis addresses the problem of predicting a single variable, in an n-by-1 response attribute variable vector $\mathbf{Y}$, from an assortment of P predictor variables (i.e., covariates)—plus a vector of ones to capture the intercept term—in an n-by-(P+1) matrix $\mathbf{X}$. One theme of conventional linear statistics is that virtually every statistical technique can be reformulated as a linear regression problem; linear regression is the workhorse of classical statistics. This contention becomes more obvious when the traditional regression technique is replaced by the general linear model, which addresses the problem of simultaneously predicting some battery of Q variables, in an n-by-Q matrix $\mathbf{Y}$, as is done with MANOVA, from those P predictor variables in matrix $\mathbf{X}$. Although multicollinearity is a concern in

Table 9.10 A comparison of $K=2$ DFA $\mathbf{W}^{-1}\mathbf{B}$ eigenvector entries with linear regression coefficients à la $\mathbf{E}_1 = c\mathbf{b}$, for some constant c

Original DFA (c = 3.659)			ESF DFA (c = 3.467)			Spatially unstructured DFA (c = 3.206)		
Attribute	**Linear regression**	**DFA**	**ESF**	**Linear regression**	**DFA**	**Residuals**	**Linear regression**	**DFA**
Population density	0.13339	0.03645	Population density	0.12491	0.03603	Population density	0.11698	0.03649
Median income	0.65074	0.17784	Median income	0.13131	0.03788	Median income	0.60930	0.19005
Female–male ratio	0.41026	0.11212	% Hispanic	0.14321	0.04131	% Public commuters	0.09319	0.02907
			% High school graduates	0.01742	0.00503	% Car transit commuters	−0.56009	−0.17470
			% Vacant houses	−0.22023	−0.06353	% High school graduates	0.00988	0.00308

regression for the variables constituting matrix **X**, the extension of vector **Y** to a matrix that includes Q variables raises issues about multicollinearity among this set of variables, too. CCA is a technique that handles this additional source of multicollinearity in the case where $P > 1$ and $Q > 1$. Accordingly, it focuses on correlations between linear combinations of variables in the two datasets. As such, it sometimes is referred to as a double-barreled PCA (geometrically, it is a PCA that then is modified by a sequence of judiciously selected axis rotations).

CCA determines matched pairs of optimal linear combinations of variables, one from each of the matrices **X** and **Y**, much like those computed with DFA and PCA or FA, but such that they are maximally related. In other words, CCA uncovers the synthetic variates from matrix **X** that are most highly correlated, linearly, to their companion synthetic variates from matrix **Y**. Just as in DFA, PCA, or FA, a structure matrix can be constructed for CCA revealing the nature and degree of relationships between the original variables and those common dimensions running across both datasets. Again, reminiscent of DFA, PCA, or FA, these synthetic variates constitute canonical correlation variate scores. The notion of canonical correlation refers to a measure of the strength of association between corresponding pairs of the synthetic variates. When $Q = 1$, this measure reduces to the multiple correlation coefficient of regression. The maximization feature is necessary because globally unique eigenfunction solutions are nonexistent (as is attested to by the use of various rotation criteria). Thus this optimization aspect of the technique enables the largest correlations to be computed. Not surprisingly, the first canonical dimension often produces a canonical correlation close or equal to one.

9.4.1 The CCA eigenfunction problem

The eigenfunction problem of CCA is constructed by partitioning the correlation matrix **R** for the combined set of variables into a submatrix $\mathbf{R}_{XX}$ for the correlations among the designated variables in matrix **X**, a submatrix $\mathbf{R}_{YY}$ for the correlations among the designated variables in matrix **Y**, and a submatrix $\mathbf{R}_{XY}$ for the cross-correlations among the variables of matrices **X** and **Y**. Because matrix **R** is symmetric (the correlation between X_p and Y_q equals the correlation between Y_q and X_p), $\mathbf{R}_{YX} = \mathbf{R}_{XY}^T$. Consequently, the CCA eigenfunction problem may be stated as follows:

$$\text{eigenvalues: } \text{Det}\left(\mathbf{R}_{XX}^{-1}\mathbf{R}_{XY}\mathbf{R}_{YY}^{-1}\mathbf{R}_{YX} - \lambda^2\mathbf{I}\right) = 0, \text{ and} \quad (9.3)$$

$$\text{eigenvectors}: \left[\begin{pmatrix} \mathbf{0} & \mathbf{R}_{XY} \\ \mathbf{R}_{YX} & \mathbf{0} \end{pmatrix} - \lambda_j \begin{pmatrix} \mathbf{R}_{XX} & \mathbf{0} \\ \mathbf{0} & \mathbf{R}_{YY} \end{pmatrix}\right]\begin{pmatrix} \mathbf{U}_j \\ \mathbf{V}_j \end{pmatrix} = \mathbf{0}, \begin{pmatrix} \mathbf{U}_j \\ \mathbf{V}_j \end{pmatrix} \neq \mathbf{0} \tag{9.4}$$

Eq. (9.3) is a reduced form of the determinant of its counterpart in Eq. (9.4). Its eigenvalue, λ^2, is the squared canonical correlation between its corresponding pair of canonical variates (CVs), computed as $\mathbf{Z}_X\mathbf{U}$ and $\mathbf{Z}_Y\mathbf{V}$ (as with PCA, the eigenvector coefficients are applied to z-score transforms of the original attribute variables). This reduced form of the eigenvalue problem highlights that if matrix **Y** is a set of spatial weights matrix eigenvectors, then the problems become

$$\text{eigenvalues}: \ \text{Det}\left(\mathbf{R}_{XX}^{-1}\mathbf{R}_{XY}\mathbf{R}_{YX} - \lambda^2\mathbf{I}\right) = 0, \ \text{and} \tag{9.5}$$

$$\text{eigenvectors}: \left[\begin{pmatrix} \mathbf{0} & \mathbf{R}_{XY} \\ \mathbf{R}_{YX} & \mathbf{0} \end{pmatrix} - \lambda_j \begin{pmatrix} \mathbf{R}_{XX} & \mathbf{0} \\ \mathbf{0} & \mathbf{I} \end{pmatrix}\right]\begin{pmatrix} \mathbf{U}_j \\ \mathbf{V}_j \end{pmatrix} = \mathbf{0}, \begin{pmatrix} \mathbf{U}_j \\ \mathbf{V}_j \end{pmatrix} \neq \mathbf{0} \tag{9.6}$$

The CVs spanning matrix **Y** are the parts of ESFs that are common to the latent mathematical dimensions of the attribute matrix **X**.

9.4.2 ESFs spanning sets of attribute variables

One MESF problem formulates matrix **Y** as the spatial weights matrix eigenvectors and matrix **X** as the georeferenced attribute variables. This formulation allows SA dimensions to be uncovered. For the data example analyzed in this chapter, matrix **Y** contains 65 positive SA eigenvectors, and matrix **X** contains 12 socioeconomic/demographic attribute variables. Table 9.11 reports features of the six uncovered latent dimensions, which have strong to marked correlations. Table 9.12 suggests that the first canonical correlation dimension may be the only truly meaningful one, with the third and

Table 9.11 A summary of CCA results for selected 2010 Texas counties socioeconomic/demographic attribute variables

Dimension	Correlation	% Variance accounted	Likelihood ratio	H_0 probability
1	0.942112	0.4173	0.00016294	<0.0001
2	0.884373	0.1897	0.00144934	<0.0001
3	0.833295	0.1201	0.00665187	<0.0001
4	0.753855	0.0696	0.02176524	<0.0001
5	0.724213	0.0583	0.05041720	<0.0001
6	0.645557	0.0378	0.10602629	0.0286

Table 9.12 The CCA structure matrix relating to SA for selected 2010 Texas counties socioeconomic/demographic attribute variables

Attribute	CCA variate (CV)					
	I	II	III	IV	V	VI
Population density	**0.6354**	0.4665	−0.2463	0.0026	−0.0218	0.1189
Female/male	0.2002	−0.0076	−0.1411	0.1258	0.0467	0.4770
% White	**−0.5265**	−0.2478	−0.4900	−0.2243	0.1183	−0.2272
% African American	**0.7671**	−0.0329	0.2460	0.4681	−0.0149	−0.0457
% Hispanic	**−0.6778**	**0.5693**	0.1863	0.2593	−0.2507	−0.0550
% Car commuters	**−0.6122**	−0.4315	0.3974	−0.0069	0.0993	0.0563
% Public transit commuters	0.2248	0.2767	0.0531	0.2059	−0.0902	−0.0137
% High school graduates	**0.5521**	−0.0966	−0.1224	−0.2521	**0.6925**	0.0301
% Unemployed	−0.0587	−0.3496	−0.3550	0.3152	−0.0501	0.1111
% Owned homes	0.3581	0.2023	0.1276	−0.4888	−0.2128	0.1114
% Vacant homes	−0.3336	−0.4617	−0.2002	0.3766	0.2198	−0.1383
Median income	0.2945	0.4439	0.3380	**−0.5852**	0.4297	−0.1365

Note: The sign pattern for a given canonical variate can be reversed because eigenvectors are unique except for a multiplicative constant of −1. A bold font denotes prominent component loadings (i.e., correlations; $|r| > 0.5$).

sixth canonical correlation dimensions failing to relate closely to any of the attribute variables. This first dimension relates to the principal Moran eigenvector, which represents a global positive SA map pattern. This type of spatial CCA merits future research.

9.5 CA and MESF

One use of CA is to extract groups of similar observations from a dataset that is composed of natural classes occurring in space. These homogeneous groups are conveniently referred to as clusters and frequently are based upon $P > 1$ variable scores measured for each of then areal units. Chapter 6 presents another form of this theme in terms of, say, local indicators of spatial association (LISA). The CA problem is more difficult than DFA, especially because in DFA the groupings are known in advance.

CA has evolved out of the taxonomic tradition in science and supplies dozens of numerical methods for objectively searching data to achieve this end. CA has manifested itself in the spatial sciences mostly through city classification and regionalization schemes. Often, researchers hold no preconceptions about the number of groups or group structure. Unfortunately, the selection of a specific clustering algorithm and threshold criteria for identifying the number of supposedly true latent groups introduces considerable subjectivity into the assessment step of CA. No single algorithm is recognized as being universally the best; and results for a given dataset commonly vary from algorithm to algorithm, sometimes quite radically. The rules of group formation utilized by each algorithm respond to different idiosyncrasies and nuances of a dataset. After all, multiple meaningful natural groups into which a set of observations can be classified may exist. In addition, clusters can be based on only those variables included in an analysis. As with other multivariate techniques, the final clusters obtained may well be sensitive to the particular choice of attribute variables included in an analysis. Therefore a spatial analyst must possess sufficient familiarity with given data, experience with the geographically distributed phenomenon under study, or an adequate conceptual framework, to distinguish between good and bad clustering results. Compactness and contiguity often are two desirable geospatial cluster features that conventional CA algorithms overlook; these two features relate to SA.

Building on the preceding ANOVA and MANOVA results, the clustering technique used for key illustrative purposes here is Ward's (1963)

Table 9.13 Hierarchical clustering results and their description of the MSA indicator variable

Algorithm output	Attribute data	ESFs	Spatially unstructured residuals
Ward's algorithm			
Number of Cluster 1 counties	37	99	127
Number of Cluster 2 counties	125	54	1
Number of Cluster 3 counties	75	56	53
Number of Cluster 4 counties	17	45	73
R^2	0.0019	0.0067	0.0081
Complete linkage algorithm			
Number of Cluster 1 counties	1	39	176
Number of Cluster 2 counties	89	48	9
Number of Cluster 3 counties	7	84	1
Number of Cluster 4 counties	157	83	68
R^2	0.0020	0.0080	0.0219
McQuitty algorithm			
Number of Cluster 1 counties	1	81	1
Number of Cluster 2 counties	1	26	1
Number of Cluster 3 counties	250	32	249
Number of Cluster 4 counties	2	115	3
R^2	0.0049	0.0008	0.0180

Note: Cluster labels 1, 2, 3, and 4 are their stepwise regression selection order.

algorithm (its accompanying dendrogram appears in Appendix A10). The Texas MSAs furnish a real-world benchmark for assessing CA results; one expectation is to uncover four regions, namely, the three major cities and the rest of the state. Table 9.13 results allude to the following two conjectures: (1) SA matters to CA results; and (2) for the limited socioeconomic/demographic data analyzed, the 2010 Texas MSAs are not a natural set of groupings (see the R^2 values in Table 9.13). Of course, these implications may change with the use of other clustering techniques or an expanded set of attribute variables. Table 9.13 also summarizes complete linkage (criterion: minimize the maximum attribute variable distance between all possible pairs of areal units, one for each of two clusters) results, which deviate even more from the a priori MSA indicator variable definition of clusters.

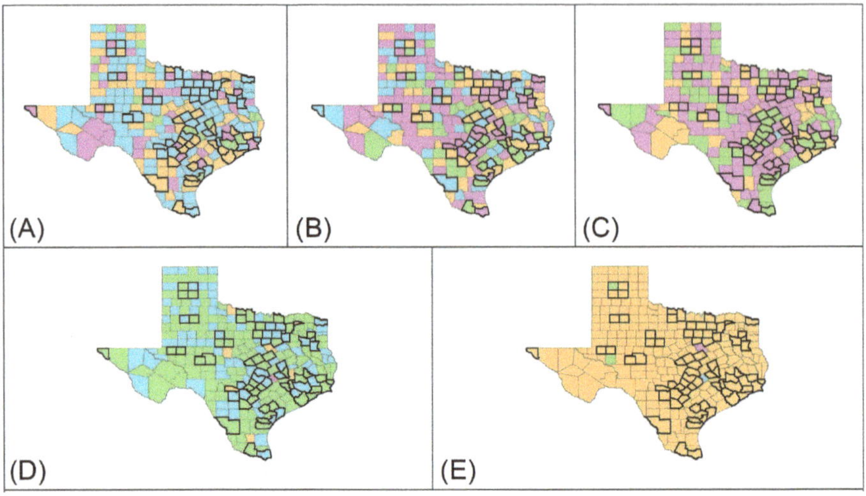

Fig. 9.1 CA classification results, with MSA counties outlined; lilac denotes Cluster #1, blue denotes Cluster #2, beige denotes Cluster #3, and green denotes Cluster #4. (A): Ward's algorithm results for the attribute data. (B): Ward's algorithm results for the ESFs. (C): Ward's algorithm results for the aspatial residuals. (D): Complete linkage algorithm results for the attribute data. (E): McQuitty algorithm results for the attribute data.

The McQuitty algorithm (criterion: minimize the weighted squared Euclidean distance between areal unit attribute variables and cluster means; each areal unit is assigned to that cluster having the nearest multivariate centroid) is a mathematically simpler form of the average linkage clustering method, with the most recent observation additions to a cluster having more weight in the distance calculations than do preceding additions.

Fig. 9.1 furnishes visualizations of selected CA taxonomic results. Fig. 9.1A–C indicates that SA helps cluster areal units (e.g., counties) into regions, whereas aspatial information tends to yield dispersed areal units in each cluster. Especially Fig. 9.1E demonstrates that some algorithms tend to differentiate only a few observations from a total set of areal units.

Including the Cartesian coordinates for the county centroids produces the results in Table 9.14. The coordinates tend to help nearby locations cluster, linking to the desired compactness and contiguity properties. However, their inclusion only marginally improves the correspondence between the prominent Cluster 1 and the MSA indicator variable. The first two Moran eigenvectors, $\mathbf{E}_1$ and $\mathbf{E}_2$, tend to correlate strongly with these coordinates.

Table 9.14 Ward's algorithm clustering results and description of the MSA indicator variable, including latitude and longitude coordinates

Algorithm output	Attribute data	ESFs	Spatially unstructured residuals
Number of Cluster 1 counties	87	87	50
Number of Cluster 2 counties	81	50	95
Number of Cluster 3 counties	70	38	1
Number of Cluster 4 counties	16	79	108
R^2	0.0102	0.0008	0.0164

Note: Cluster labels 1, 2, 3, and 4 are their stepwise regression selection order.

9.6 Summary

This chapter presents an overview of traditional linear multivariate statistics applied to geospatial data, with an emphasis on SA, its data analytic impacts, and its representation by ESFs, supplementing more comprehensive generalized linear model content (see Chapter 5) with general linear model content. One conspicuous finding is that SA matters in multivariate as well as univariate georeferenced data analyses and hence also needs to be recognized and efficiently and effectively accounted for and handled in these former analyses. This chapter's discussion addresses the basic multivariate statistical techniques (i.e., principal components analysis, factor analysis, multivariate analysis of variance, discriminant function analysis, canonical correlation analysis, and cluster analysis), beyond multiple linear regression, which is the topic of much of the rest of this book. This chapter examines empirical contrasts between a set of original attribute variables, some of which have been subjected to a Box–Cox logarithmic transformation to better align them with a bell-shaped frequency distribution, and datasets comprising their latent spatially structured (i.e., ESF) and spatially unstructured (i.e., linear regression residuals) components. General conclusions include that SA impacts upon multicollinearity as well as multiple testing—such as that embedded in MANOVA and DFA—sometimes in a substantial way. Mathematical dimensions spanning either a single dataset or pairs of datasets are a function of SA. Differences between and differentiation of regional groupings of areal units are functions of SA. More natural areal unit clusterings of georeferenced data also are a function of SA. Although the SA literature focuses on its impact upon variance, this chapter demonstrates that it also impacts upon correlation coefficients, confirming the importance of work by Clifford, Richardson, and Hémon (1989), Haining (1991), and Vallejos (2008). Future research needs to address impacts of SA on contemporary nonlinear multivariate statistics. Although linear regression is the workhorse of traditional linear statistics, nonlinear regression is the workhorse of current state-of-the-art spatial statistics and spatial econometrics.

Appendix 9.A A dendogram from Ward's algorithm for original attribute data

Fig. 9.A.1 portrays the complete dendrogram for n = 254 Texas counties, furnishing the overall hierarchical classification structure.

Fig. 9.A.2 furnishes magnified horizontal sections (from left to right) of the Fig. 9.A.1 dendrogram in order to make legible each county identification number.

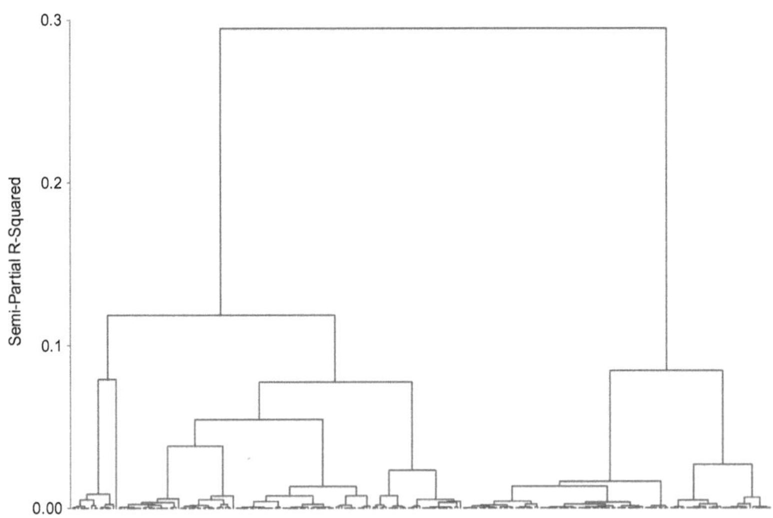

Fig. 9.A.1 The complete dendrogram for a clustering of Texas counties based upon Ward's algorithm.

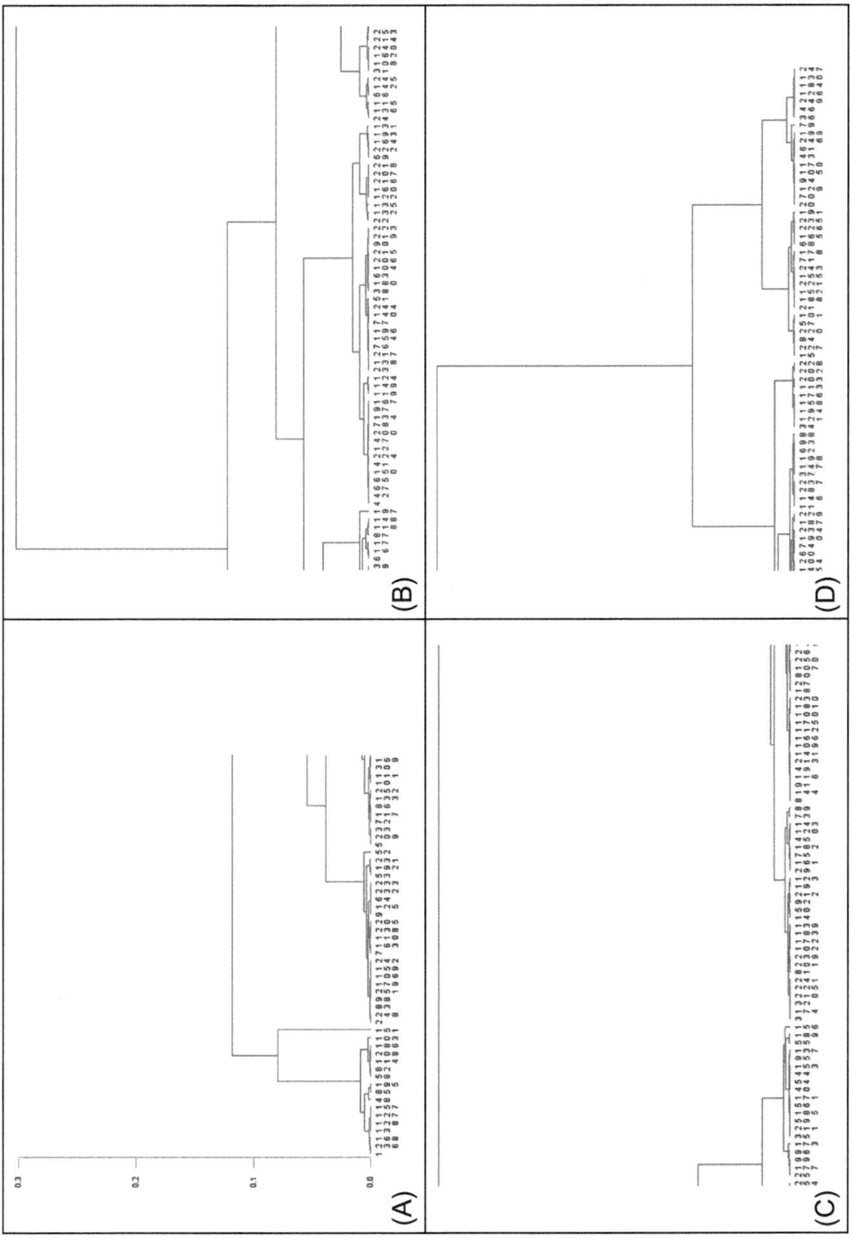

Fig. 9.A.2 Magnified horizontal sections of the Fig. 9.A.1 dendrogram. (A) Left-hand side. (B) Left-middle. (C) Right-middle. (D) Right hand side.

Appendix 9.B Multivariate statistical analysis R code

```
## set a working directory
#setwd("C:/IntroESF")
setwd("D:/proj/IntroESF/Ch9")

## load libraries and functions
library(rgdal)
library(spdep)
library(psych)
source("D:/proj/IntroESF/functions.R")

## read data and transform variables
tx.pop <- read.csv("TX-space-time-population.csv")
tx.poly <- readOGR(dsn=getwd(), layer="TX_counties_spcs",
GDAL1_integer64_policy=TRUE)

## socio-economic variable transformation
tx.df <- data.frame(pd=log(tx.poly$Pop_Total / tx.poly$ALAND))
tx.df$fmratio <- tx.poly$Pop_Female / tx.poly$Pop_Male
tx.df$pctw <- tx.poly$Pop_White / tx.poly$Pop_Total
tx.df$pctaa <- log(100 * tx.poly$Pop_Black / tx.poly$Pop_Total +
0.6)
tx.df$pcthisp <- log(100 * tx.poly$Pop_Hisp / tx.poly$Pop_Total)
tx.df$pctccar <- log(100 * tx.poly$Com_Car / tx.poly$Com_SumTra)
tx.df$pctcpub <- log(100 * tx.poly$Com_Public / tx.poly$Com_SumTra
+ 0.002)
tx.df$pcthied <- 100 * tx.poly$Edu_HSD_hi / tx.poly$Edu_25Olde
tx.df$pctunemp <- 100 * tx.poly$Uemp_NotLa / tx.poly$Uemp_16Old
tx.df$pcthowner <- 100 * tx.poly$House_Owne / tx.poly$Housing_To
tx.df$pcthvac <- log(100 * tx.poly$House_Vaca / tx.poly
$Housing_To)
tx.df$lnincome <- log(tx.poly$Inc_MedHH)

## generate eigenvectors
tx.nb <- poly2nb(tx.poly, queen=FALSE)
n <- length(tx.nb)
B <- nb2mat(tx.nb, style="B")
M <- diag(n) - matrix(1, n, n)/n
MBM <- M %*% B %*% M
eig <- eigen(MBM, symmetric=T)
```

```
EV <- as.data.frame(eig$vectors)
colnames(EV) <- paste("EV",1:n, sep="")
EV <- as.data.frame(EV[, eig$values/eig$values[1] > 0.25])

## correlation for 3 variables
# correction among attributes
tx.3 <- tx.df[, c("pd", "pctunemp", "lnincome")]
round(cor(tx.3), 4)    # Table 9.1a
tx3.pca <- principal(tx.3, 3, rotate="none")
tx3.pca$values[1]/tx3.pca$values[3]

# MESF for the 3 variables
lmf.pd <- lm(tx.df$pd ~ ., data=EV)
esf.pd <- stepwise(lmf.pd , lm(tx.df$pd ~ 1, data=EV),
verbose=FALSE)
lmf.unemp <- lm(tx.df$pctunemp ~ ., data=EV)
esf.unemp <- stepwise(lmf.unemp, lm(tx.df$pctunemp ~ 1, data=EV),
verbose=FALSE)
lmf.inc <- lm(tx.df$lnincome ~ ., data=EV)
esf.inc <- stepwise(lmf.inc, lm(tx.df$lnincome ~ 1, data=EV),
verbose=FALSE)

# correlation among aspatial compoents
asp.3 <- cbind(resid(esf.pd) + coef(esf.pd)[1],
               resid(esf.unemp) + coef(esf.unemp)[1],
               resid(esf.inc) + coef(esf.inc)[1])
round(cor(asp.3), 4)   # Table 9.1b
asp3.pca <- principal(asp.3, 3, rotate="none")
asp3.pca$values[1]/asp3.pca$values[3]

# selected EVs
ev.pd <- names(coef(esf.pd)[-1])
ev.unemp <- names(coef(esf.unemp)[-1])
ev.inc <- names(coef(esf.inc)[-1])

# find common EVs
ev.c <- Reduce(intersect, list(ev.pd, ev.unemp, ev.inc))

# run ESF without the common EVs
ev.nc <- setdiff(colnames(EV), ev.c)
lmf.pd.nc <- lm(tx.df$pd ~ ., data=EV[,ev.nc])
```

```
esf.pd.nc <- stepwise(lmf.pd.nc, lm(tx.df$pd ~ 1, data=EV[,ev.
nc]), verbose=FALSE)
lmf.unemp.nc <- lm(tx.df$pctunemp ~ ., data=EV[,ev.nc])
esf.unemp.nc <- stepwise(lmf.unemp.nc, lm(tx.df$pctunemp ~ 1,
data=EV[,ev.nc]), verbose=FALSE)
lmf.inc.nc <- lm(tx.df$lnincome ~ ., data=EV[,ev.nc])
esf.inc.nc <- stepwise(lmf.inc.nc, lm(tx.df$lnincome ~ 1, data=EV
[,ev.nc]), verbose=FALSE)

# selected EVs for the updated ESFs
ev.pd.nc <- names(coef(esf.pd.nc)[-1])
ev.unemp.nc <- names(coef(esf.unemp.nc)[-1])
ev.inc.nc <- names(coef(esf.inc.nc)[-1])

# unique eigenvectors for the updated ESFs
evu.pd <- setdiff(ev.pd.nc, union(ev.unemp.nc, ev.inc.nc))
evu.unemp <- setdiff(ev.unemp.nc, union(ev.pd.nc, ev.inc.nc))
evu.inc <- setdiff(ev.inc.nc, union(ev.pd.nc, ev.unemp.nc))

# correlation among aspatial compoents + unique eigenvectors
pd.u <- lm(tx.df$pd ~ ., data=EV[,evu.pd])
unemp.u <- lm(tx.df$pctunemp ~ ., data=EV[,evu.unemp])
inc.u <- lm(tx.df$lnincome ~ ., data=EV[,evu.inc])
au3 <- cbind(resid(esf.pd) + fitted(pd.u),
             resid(esf.unemp) + fitted(unemp.u),
             resid(esf.inc) + fitted(inc.u))
round(cor(au3), 4)  #Table 9.1c
au3.pca <- principal(au3, 3, rotate="none")
au3.pca$values[1]/au3.pca$values[3]

# correlation among aspatial compoents + common eigenvectors
pd.c <- lm(tx.df$pd ~ ., data=EV[,ev.c])
unemp.c <- lm(tx.df$pctunemp ~ ., data=EV[,ev.c])
inc.c <- lm(tx.df$lnincome ~ ., data=EV[,ev.c])
ac3 <- cbind(resid(esf.pd) + fitted(pd.c),
             resid(esf.unemp) + fitted(unemp.c),
             resid(esf.inc) + fitted(inc.c))
round(cor(ac3), 4)  #Table 9.1d
ac3.pca <- principal(ac3, 3, rotate="none")
ac3.pca$values[1]/ac3.pca$values[3]
```

```
## PCA for Texas Population
library(psych)
tx.lpd <- log(tx.pop[,2:8]/tx.pop$area)
lpd.pca <- principal(tx.lpd, 7, rotate="none")
print(lpd.pca$loadings, cutoff=0.001) # Table 9.2

n.year <- 7
lpd.esf <- matrix(0, nrow=n, ncol=n.year)
lpd.asp <- matrix(0, nrow=n, ncol=n.year)
colnames(lpd.esf) <- paste("y", 2010:2016, sep="")
colnames(lpd.asp) <- paste("y", 2010:2016, sep="")

for (i in 1:n.year) {
     lm.full <- lm(tx.lpd[,i] ~ ., data=EV)
     sf.i <- stepwise(lm.full, lm(tx.lpd[,i] ~ 1, data=EV),
     verbose=FALSE)
     lpd.esf[,i] <- sf.i$fitted.values
     lpd.asp[,i] <- coef(sf.i)[1] + resid(sf.i)
}

lpd.esf.pca <- principal(lpd.esf, 7, rotate="none")
print(lpd.esf.pca$loadings, cutoff=0.001) # Table 9.3 Part I

lpd.asp.pca <- principal(lpd.asp, 7, rotate="none")
print(lpd.asp.pca$loadings, cutoff=0.001) # Table 9.3 Part II

## MESF for the 12 socio-economic variables
n.var <- dim(tx.df)[2]
esf.mat <- matrix(0, nrow=n, ncol=n.var)
asp.mat <- matrix(0, nrow=n, ncol=n.var)
var.names <- c("pd", "fmratio", "pctw", "pctaa", "pcthisp",
"pctccar",
               "pctcpub", "pcthied", "pctunemp", "pcthowner",
               "pcthvac", "lnincome")
colnames(esf.mat) <- var.names
colnames(asp.mat) <- var.names

for (i in 1:n.var) {
     lm.full <- lm(tx.df[,i] ~ ., data=EV)
     sf.i <- stepwise(lm.full, lm(tx.df[,i] ~ 1, data=EV),
     sle=0.1, sls=0.1001, verbose=FALSE)
```

```
    esf.mat[,i] <- sf.i$fitted.values
    asp.mat[,i] <- sf.i$coefficients[1] + residuals(sf.i)
}

## PCA
tx.pca <- principal(tx.df, 7, rotate="none")
print(tx.pca$loadings, cutoff=0.001) # Table 9.4

esf.pca <- principal(esf.mat, 7, rotate="none")
print(esf.pca$loadings, cutoff=0.001) # Table 9.5 Part I

asp.pca <- principal(asp.mat, 7, rotate="none")
print(asp.pca$loadings, cutoff=0.001) # Table 9.5 Part II

## FA
tx.fa <- principal(tx.df, 6, rotate="varimax")
print(tx.fa$loadings, cutoff=0.001)  # Table 9.6

esf.fa <- principal(esf.mat, 4, rotate="varimax")
print(esf.fa$loadings, cutoff=0.001) # Table 9.7 Part I

asp.fa <- principal(asp.mat, 7, rotate="varimax")
print(asp.fa$loadings, cutoff=0.001) # Table 9.7 Part II

## ANOVA & MANOVA
mas.f <- as.factor(tx.poly$MSA)

f.av <- function (x, z){
    anova(lm(x ~ z))[[5]][1]
}

f.lev <- function(x, z){
    leveneTest.square(x, z)[[3]][1]
}

tx.p <- apply(tx.df, 2, FUN=f.av, z=mas.f)
tx.lev <- apply(tx.df, 2, FUN=f.lev, z=mas.f)

esf.p <- apply(esf.mat, 2, FUN=f.av, z=mas.f)
esf.lev <- apply(esf.mat, 2, FUN=f.lev, z=mas.f)

asp.p <- apply(asp.mat, 2, FUN=f.av, z=mas.f)
asp.lev <- apply(asp.mat, 2, FUN=f.lev, z=mas.f)
```

```
round(cbind(tx.p, tx.lev, esf.p, esf.lev, asp.p, asp.lev),4) #
Table 9.8

summary(manova(as.matrix(tx.df)~mas.f))
summary(manova(esf.mat~mas.f))
summary(manova(asp.mat~mas.f))

## Discriminant Analysis
library(MASS)
library(klaR)
tx.gw <- greedy.wilks(mas.f ~ ., data = tx.df, niveau = 0.1)
tx.gw    # Table 9.9a

esf.df <- as.data.frame(esf.mat)
esf.gw <- greedy.wilks(mas.f ~ ., data = esf.df, niveau = 0.1)
esf.gw          # Table 9.9b

asp.df <- as.data.frame(asp.mat)
asp.gw <- greedy.wilks(mas.f ~ ., data = asp.df, niveau = 0.1)
asp.gw          # Table 9.9c

fml.tx <- update(tx.gw$formula, tx.poly$MSA ~ .)
tx.lm <- lm(fml.tx, data=tx.df)
summary(tx.lm)  # Table 9.10a

fml.esf <- update(esf.gw$formula, tx.poly$MSA ~ .)
esf.lm <- lm(fml.esf, data=esf.df)
summary(esf.lm)  # Table 9.10b

fml.asp <- update(asp.gw$formula, tx.poly$MSA ~ .)
asp.lm <- lm(fml.asp, data=asp.df)
summary(asp.lm)  # Table 9.10c

## Canonical Correlation Analysis
library(CCA)
library(CCP)
tx.cc <- cc(EV, tx.df)
cc.r <- tx.cc$cor
print(round(cc.r,4))            # Table 9.11 correlation

cc.ev <- cc.r^2/(1-cc.r^2)
round(cc.ev/sum(cc.ev),4)       # Table 9.11 % of variance
```

```
# Calculate p-values with Wilk's lambda
p.asym(cc.r, n, dim(EV)[2], dim(tx.df)[2])      # Table 9.11 Test

round(tx.cc$scores$corr.Y.yscores[,1:6],4)      # Table 9.12

## Cluster Analysis
# for the attributes (Table 9.13)
tx.dist <- dist(scale(tx.df))^2
tx.ward <- hclust(tx.dist, method="ward.D2")
tx.ward.g <- cutree(tx.ward, k=4)
table(tx.ward.g)
plot(tx.ward)
rect.hclust(tx.ward, k=4, border="red")

tx.comp <- hclust(tx.dist, method="complete")
tx.comp.g <- cutree(tx.comp, k=4)
table(tx.comp.g)

tx.mq <- hclust(tx.dist, method="mcquitty")
tx.mq.g <- cutree(tx.mq, k=4)
table(tx.mq.g)

# for ESFs (Table 9.13)
esf.dist <- dist(scale(esf.mat))^2

esf.ward <- hclust(esf.dist, method="ward.D2")
esf.ward.g <- cutree(esf.ward, k=4)
table(esf.ward.g)

esf.comp <- hclust(esf.dist, method="complete")
esf.comp.g <- cutree(esf.comp, k=4)
table(esf.comp.g)

esf.mq <- hclust(esf.dist, method="mcquitty")
esf.mq.g <- cutree(esf.mq, k=4)
table(esf.mq.g)

# for unstructured residuals (Table 9.13)
asp.dist <- dist(scale(asp.mat))^2

asp.ward <- hclust(asp.dist, method="ward.D2")
asp.ward.g <- cutree(asp.ward, k=4)
table(asp.ward.g)

asp.comp <- hclust(asp.dist, method="complete")
```

```
asp.comp.g <- cutree(asp.comp, k=4)
table(asp.comp.g)

asp.mq <- hclust(asp.dist, method="mcquitty")
asp.mq.g <- cutree(asp.mq, k=4)
table(asp.mq.g)

# with coordinates (Table 9.14)
c.x <- as.numeric(as.vector(tx.poly$INTPTLON))
c.y <- as.numeric(as.vector(tx.poly$INTPTLAT))

# MESF for the longitudes and latitudes
lm.full <- lm(c.x ~ ., data=EV)
sf.x <- stepwise(lm.full, lm(c.x ~ 1, data=EV), sle=0.1,
sls=0.1001, verbose=FALSE)
esf.x <- sf.x$fitted.values
asp.x <- sf.x$coefficients[1] + residuals(sf.x)

lm.full <- lm(c.y ~ ., data=EV)
sf.y <- stepwise(lm.full, lm(c.y ~ 1, data=EV), sle=0.1,
sls=0.1001, verbose=FALSE)
esf.y <- sf.y$fitted.values
asp.y <- sf.y$coefficients[1] + residuals(sf.y)

txc.df <- cbind(tx.df, c.x, c.y)
txc.dist <- dist(scale(txc.df))^2

txc.ward <- hclust(txc.dist, method="ward.D2")
txc.ward.g <- cutree(txc.ward, k=4)
table(txc.ward.g)   # Table 9.14, column 1

esfc.df <- cbind(esf.df, esf.x, esf.y)
esfc.dist <- dist(scale(esfc.df))^2

esfc.ward <- hclust(esfc.dist, method="ward.D2")
esfc.ward.g <- cutree(esfc.ward, k=4)
table(esfc.ward.g)   # Table 9.14, column 2

aspc.df <- cbind(asp.df, asp.x, asp.y)
aspc.dist <- dist(scale(aspc.df))^2

aspc.ward <- hclust(aspc.dist, method="ward.D2")
aspc.ward.g <- cutree(aspc.ward, k=4)
table(aspc.ward.g)   # Table 9.14, column 3
```

References

Clifford, P., Richardson, S., & Hémon, D. (1989). Assessing the significance of the correlation between two spatial processes. *Biometrics*, *45*, 123–134.

Griffith, D., & Amrhein, C. (1997). *Multivariate statistical analysis for geographers*. Englewood Cliffs, NJ: Prentice Hall.

Haining, R. (1991). Bivariate correlation with spatial data. *Geographical Analysis*, *23*(3), 210–227.

Johnson, R., & Wichern, D. (2007). *Applied multivariate statistical analysis* (6th ed.). Upper Saddle River, NJ: Pearson-Prentice Hall.

Montgomery, D., Peck, E., & Vining, G. (2001). *Introduction to linear regression analysis* (3rd ed.). New York: Wiley.

Pearson, K. (1901). On lines and planes of closest fit to systems of points in space. *Philosophical Magazine*, *2*(11), 559–572 (Series 6).

Revelle, W., & Rocklin, T. (1979). Very simple structure: an alternative procedure for estimating the optimal number of interpretable factors. *Multivariate Behavioral Research*, *14*, 403–414. https://doi.org/10.1207/s15327906mbr1404_2.

Tatsuoka, M., & Lohnes, P. (1988). *Multivariate analysis* (2nd ed.). New York: Macmillan.

Vallejos, R. (2008). Assessing the association between two spatial or temporal sequences. *Journal of Applied Statistics*, *35*(12), 323–343.

Ward, J. (1963). Hierarchical grouping to optimize an objective function. *Journal of the American Statistical Association*, *58*, 236–244.

Welch, B. (1951). On the comparison of several mean values: an alternative approach. *Biometrika*, *38*, 330–336.

CHAPTER 10

Concluding comments: Toy dataset implementation demonstrations

Chapter outline

A conclusion for this book could take on many different forms, such as a chapter-by-chapter summary (this appears in the Preface as well as the Epilogue, but in two different forms), or the sketching of a future research agenda (a selective version of such a program appears in the Epilogue). Because the focus of this book is Moran eigenvector spatial filtering (MESF), the chosen form for this chapter is screenshots of output for a toy dataset (Kamp and Partee, 1995) example—in keeping with the toy examples appearing in Sections 7.5 and 8.3—employing the Spatial Analysis using ArcGIS Engine and R (SAAR) v.1.0.7 software discussed in Chapter 4 (Koo, Chun, and Griffith, 2018). This content facilitates replicability and supports repeatability for selected results like those appearing in Chapters 1–3, 5, and 6.

10.1 The toy example: A Dallas–Fort Worth metroplex county geographic resolution dataset

Fig. 2.4 portrays census tract results for the Dallas–Fort Worth (DFW) metropolitan region. This same region, a United States (US) Census metropolitan

Spatial Regression Analysis Using Eigenvector Spatial Filtering
https://doi.org/10.1016/B978-0-12-815043-6.00009-4

statistical area (MSA) partitioned into 12 counties,[1] constitutes a toy dataset for illustrative purposes. Table 10.1 reports the attributes used in analyses summarized in this chapter. Fig. 10.1 portrays selected geographic distributions that are comparable to those appearing in previous chapters. Population density (Fig. 10.1B) depicts a Poisson random variable (RV) for counts, with a log-area offset variable. Percentage Hispanic (Fig. 10.1C) depicts a binomial RV for constrained counts (Hispanic counts that are a subset of total population counts). Box–Cox transformed versions of the density, as well as the percentage itself, may be treated as normal RV approximations. The binary 0–1 indicator variable depicts a Bernoulli RV (for Bernoulli/logistic regression), indicating whether a county's percentage of population in poverty is above the metroplex average; these data are a binomial RV with a constraint on counts of one.

Table 10.2 presents binary spatial weights matrices for the two common adjacency definitions: rook and queen. The SAAR software constructs these matrices from a .shp file.

10.2 The setup

SAAR (Fig. 10.2B portrays its icon) has some functionalities in common with ArcMap, primarily because it is built upon ArcGIS Engine. Its abridged version, the ESF Tool (Fig. 10.2C portrays its icon), with limited functionalities, is not. Fig. 10.2D portrays the splash screen that opens by clicking on the SAAR icon. Fig. 10.2A depicts the opening main window of SAAR. This illustration focuses on the ***Moran Scatterplot*** option in the ***Explore*** pull-down menu and the ***Eigenvector Spatial Filtering*** option in the ***Regression*** pull-down menu. The demonstrations in this chapter are for pure spatial autocorrelation (SA) cases, without substantive covariates [i.e., intercept-only eigenvector spatial filter (ESF) terms].

10.3 Moran scatterplots

The two normal RV approximations are transformed population density (see Section 3.2) and percentage of Hispanics (which is untransformed).[2]

[1] Some metropolitan demarcations of the DFW metroplex include Hood and/or Grayson Counties, which have not been included here.

[2] The data file contains the original data as well as the transformed data (tr_pd), and area as well as log-area (LN_area). These transformed data can be calculated utilizing the ***Data*** pull-down menu, using ***Add Field***, ***Field Calculator***, and ***Field Transformation***.

Table 10.1 The selected data for the toy example

					Population			
					Total			
County	Id	Latitude	Longitude	Area	2010	2016	2010 Hispanic	2010 Poverty
Collin	1	33.1947365	−96.5797070	217.89	782,341	939,585	125,023	65,433 (0)
Dallas	2	32.7669866	−96.7784238	226.03	2,368,139	2,574,984	950,159	467,347 (1)
Denton	3	33.2051320	−97.1211403	227.53	662,614	806,180	132,267	61,925 (0)
Ellis	4	32.3468783	−96.7969403	242.33	149,610	168,499	37,587	18,294 (0)
Henderson	5	32.2116334	−95.8534183	226.29	78,532	79,901	9035	15,311 (1)
Hood	6	32.4301490	−97.8316766	108.95	51,182	56,857	5745	6053 (0)
Hunt	7	33.1232921	−96.0842024	217.64	86,129	92,073	12,438	16,502 (1)
Johnson	8	32.3798207	−97.3649587	187.72	150,934	163,274	29,353	18,846 (0)
Kaufman	9	32.5989440	−96.2883777	202.20	103,350	118,350	19,544	14,628 (0)
Parker	10	32.7770964	−97.8059049	234.00	116,927	129,441	13,172	12,874 (0)
Rockwall	11	32.8986766	−96.4044017	32.92	78,337	93,978	13,714	5156 (0)
Tarrant	12	32.7721191	−97.2912241	223.67	1,809,034	2,016,872	513,740	284,626 (1)

Note: The binary poverty variable indicates that a county percentage in poverty is greater than the MSA-wide percentage, the expected value for a random mixture.

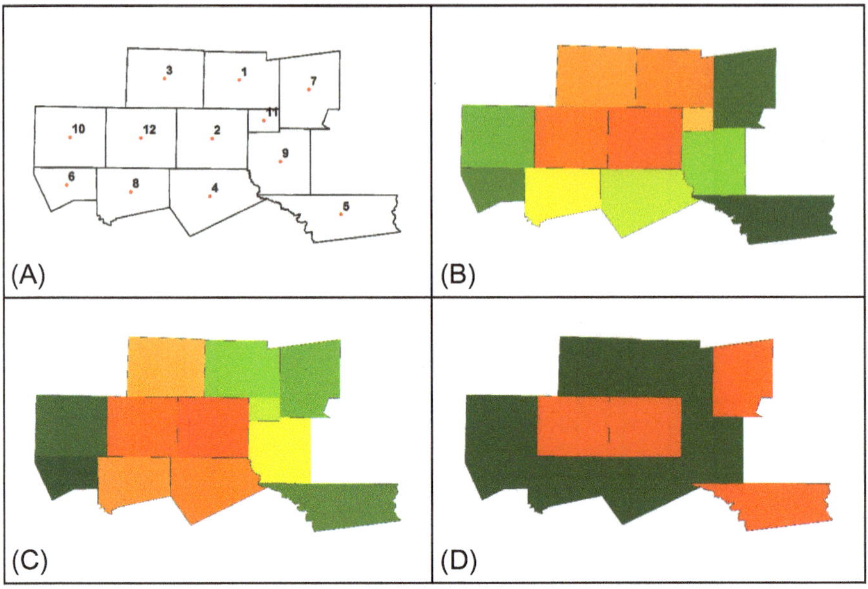

Fig. 10.1 The quantity of a variable increases from *dark green* to *yellow* to *dark red*. (A) Spatial structure of the region (*red denotes* geometric county centroids). (B) Box–Cox transformed population density. (C) Percentage population that is Hispanic. (D) Poverty percentage above (*red*) and below (*green*) the MSA-wide average.

The former has a Moran coefficient (MC) of 0.293 and a Geary ratio (GR) of 0.685, whereas the latter has MC = 0.104 and GR = 0.909. Fig. 10.3 portrays their respective Moran scatterplots. The *straight lines* are trend lines whose slopes are proportional to their respective global MC values. The axes highlight local indicators of SA (LISA; see Section 6.4.1); in counterclockwise order, beginning with the first quadrant: high–high (HH), low–high (LH), low–low (LL), and high–low (HL).

10.4 Normal approximation regression: The spatial linear regression specification

Both the Box–Cox transformed population density and the percentage of Hispanic population can be subjected to a spatial linear regression analysis. The illustrative analysis here employs the simultaneous autoregressive (SAR) specification. Fig. 10.4A portrays the opened dialogue window, which requires entering the .shp file for the DFW example. As for the attribute variables, the context menu on a layer in the main window allows an attribute table to be opened and variable transformations calculated. The Box–Cox

Table 10.2 Binary 0–1 spatial weights matrices for the DFW metroplex (see Fig. 10.1A).

Rook adjacency													Queen adjacency												
County	**1**	**2**	**3**	**4**	**5**	**6**	**7**	**8**	**9**	**10**	**11**	**12**	**County**	**1**	**2**	**3**	**4**	**5**	**6**	**7**	**8**	**9**	**10**	**11**	**12**
1	0	1	1	0	0	0	1	0	0	0	1	0	**1**	0	1	1	0	0	0	1	0	0	0	1	0
2	1	0	1	1	0	0	0	0	1	0	1	1	**2**	1	0	1	1	0	0	0	0	1	0	1	1
3	1	1	0	0	0	0	0	0	0	0	0	1	**3**	1	1	0	0	0	0	0	0	0	0	0	1
4	0	1	0	0	1	0	0	1	1	0	0	1	**4**	0	1	0	0	1	0	0	1	1	0	0	1
5	0	0	0	1	0	0	0	0	1	0	0	0	**5**	0	0	0	1	0	0	0	0	1	0	0	0
6	0	0	0	0	0	0	0	1	0	1	0	0	**6**	0	0	0	0	0	0	0	1	0	1	0	0
7	1	0	0	0	0	0	0	0	1	0	1	0	**7**	1	0	0	0	0	0	0	0	1	0	1	0
8	0	0	0	1	0	1	0	0	0	1	0	1	**8**	0	0	0	1	0	1	0	0	0	1	0	1
9	0	1	0	1	1	0	1	0	0	0	1	0	**9**	0	1	0	1	1	0	1	0	0	0	1	0
10	0	0	0	0	0	1	0	1	0	0	0	1	**10**	0	0	0	0	0	1	0	1	0	0	0	1
11	1	1	0	0	0	0	1	0	1	0	0	0	**11**	1	1	0	0	0	0	1	0	1	0	0	0
12	0	1	1	1	0	0	0	1	0	1	0	0	**12**	0	1	1	1	0	0	0	1	0	1	0	0

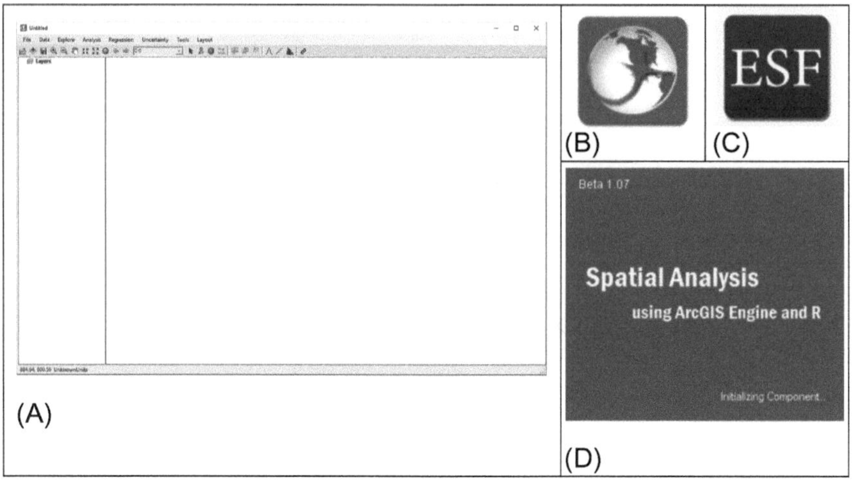

Fig. 10.2 Opening SAAR. (A) The opening main window. (B) The SAAR icon. (C) The ESF Tool icon. (D) The SAAR splash screen.

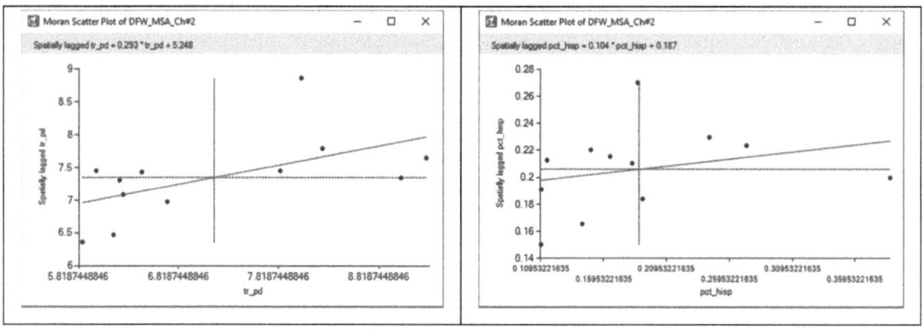

(A) (B)

Fig. 10.3 Moran scatterplots. (A) Box–Cox transformed population density. (B) Percentage of Hispanic population.

transformation used in this example is the one determined for the statewide Texas data and is denoted by tr_pd; this variable is selected as the ***Dependent Variable***. The selected ***Spatial Weight Matrix*** is for the queen's definition of adjacency. Producing a pure SA description uses only the default ***intercept*** variable already appearing in the list box for independent variables. The SAR model is selected, as is ***Eigen***, to calculate the Jacobian term. Finally, the ***Save regression residuals*** box is checked so that residuals can be mapped. Clicking on the ***Run*** button executes this analysis. Fig. 10.4C portrays the generated output: the intercept is 6.99329, and λ (i.e., the SA parameter) is 0.52353. Meanwhile, right-clicking on the layer

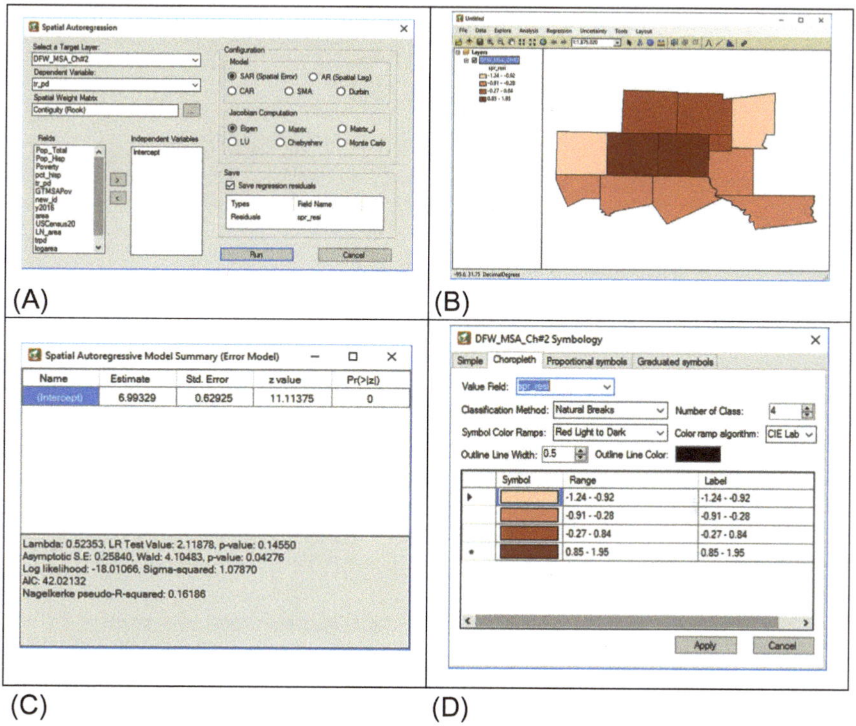

Fig. 10.4 An SAR analysis of transformed DFW population density. (A) The analysis setup window. (B) The map of residuals window. (C) The regression output window. (D) The choropleth mapping window.

opens the context window from which ***Symbology*** can be selected, which in turn has a ***Choropleth*** tab (Fig. 10.4D) whose completion allows residuals to be mapped. Fig. 10.4B portrays this map.

Fig. 10.5 reports the results from replicating this analysis but with a rook's definition of adjacency and percentage of Hispanic population (pct_hisp). Here the intercept is 0.18211, and λ is 0.24203.

10.5 Poisson regression: The MESF specification

Population density also can be described with total population as a Poisson RV and log-area as an offset. This specification captures the true nature of counts data and suggests that a normal approximation should involve a logarithmic Box–Cox power transformation (which it does; see Section 3.2). The illustrative analysis here employs the MESF generalized linear model (GLM) specification. As before, Fig. 10.6A portrays the opened dialogue

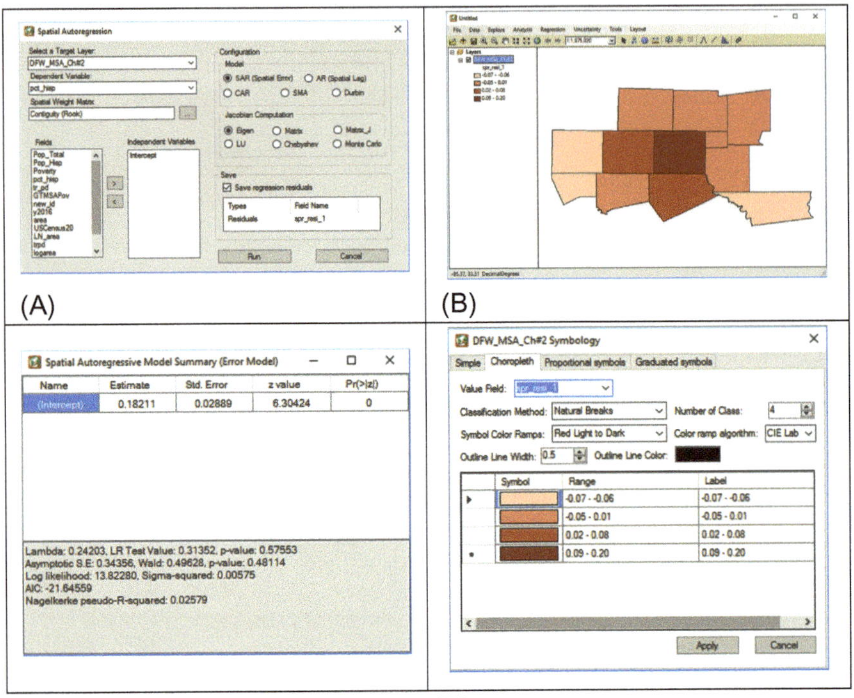

(C) (D)

Fig. 10.5 An SAR analysis of DFW's percentage of Hispanic population. (A) The analysis setup window. (B) The map of residuals window. (C) The regression output window. (D) The choropleth mapping window.

window, which requires entering the .shp file for the DFW example. Next, ***Eigenvector Spatial Filtering*** is selected from the ***Regression*** pull-down menu. Poisson is selected from the ***Family*** option, Pop_Total is inserted as the ***Dependent Variable***, and LN_area is inserted as the ***Offset***. The selected ***Spatial Weight Matrix*** is for the rook's definition of adjacency. Producing a pure SA description uses only the default ***intercept*** variable already appearing in the list box for independent variables. ***Eigenvalues over the principal >*** is selected, retaining its default value of 0.25, and ***Direction*** is set to ***Positive Only*** (the analysis is restricted to positive SA). Finally, the ***Show coefficients of selected eigenvectors*** and ***Save residuals and spatial filter*** boxes are checked, for mapping purposes; the first box is checked only so that the eigenvector results are not hidden. Clicking on the ***Run*** button executes this analysis. Fig. 10.6C portrays the generated output: the intercept is 7.32214, and the first three eigenvectors (EV3, EV2, and EV1, acknowledging their selection order) are selected to describe SA.

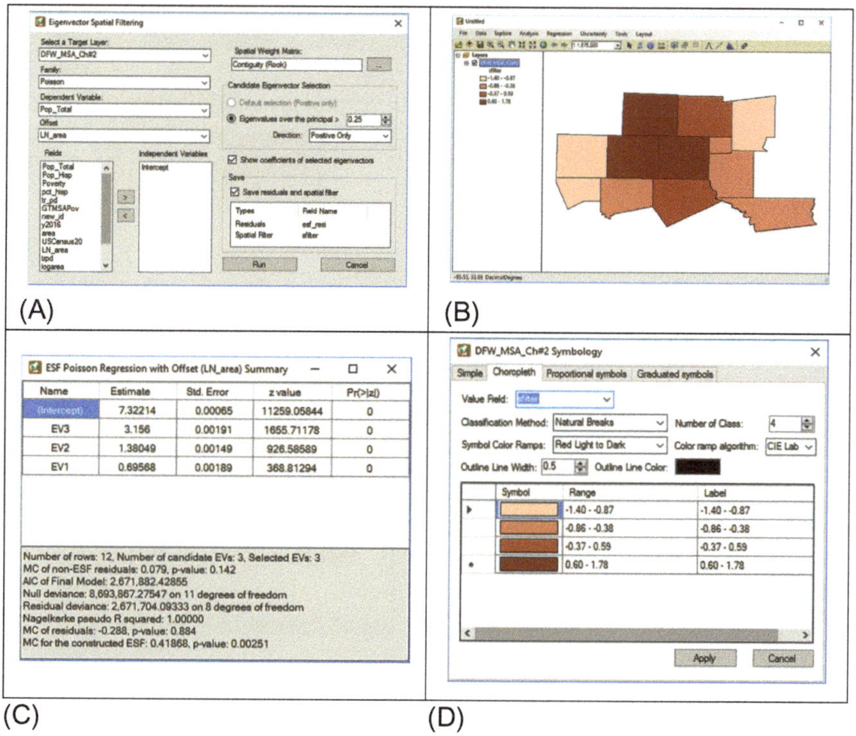

Fig. 10.6 A MESF analysis of transformed DFW population density. (A): The analysis setup window. (B) The ESF map window. (C) The regression output window. (D) The choropleth mapping window.

Meanwhile, right-clicking on the layer opens a window from which ***Symbology*** can be selected, which in turn has a ***Choropleth*** tab (Fig. 10.6D) whose completion allows the constructed ESF to be mapped. Fig. 10.6B portrays this map.

10.6 Binomial regression: The MESF specification

The percentage of Hispanic population also can be analyzed such that Hispanic population counts constitute a binomial RV with draws from their corresponding total population counts (e.g., number of trials). This representation captures the true nature of these data (assuming normality introduces some degree of specification error). Again, the illustrative analysis here employs the MESF GLM specification. As before, Fig. 10.7A portrays the opened dialogue window, which requires entering the .shp file for the

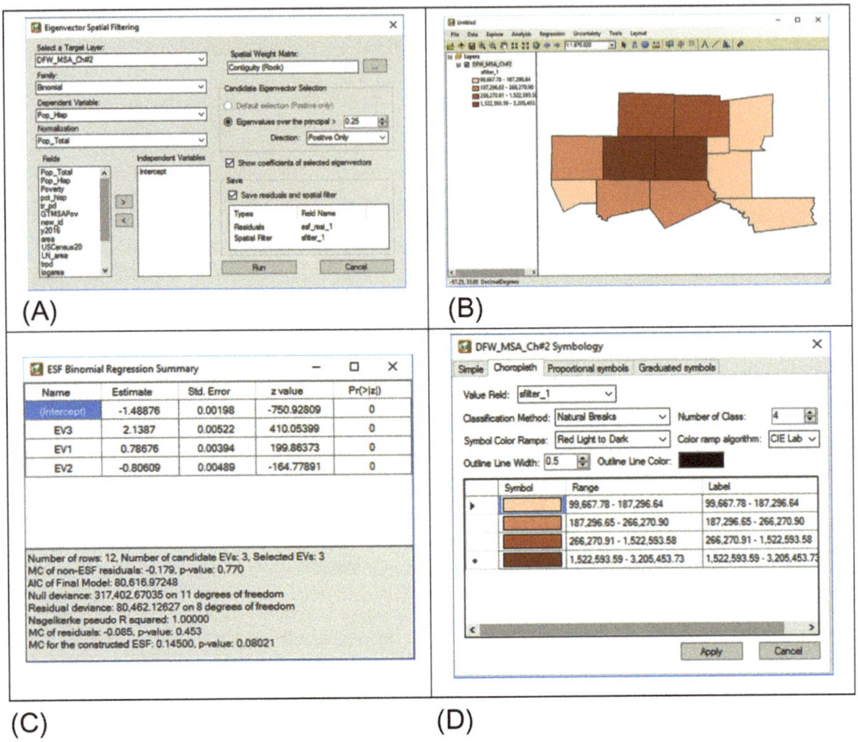

Fig. 10.7 A MESF analysis of DFW's percentage of Hispanic population. (A) The analysis set up window. (B) The ESF map window. (C) The regression output window. (D) The choropleth mapping window.

DFW example. Next, ***Eigenvector Spatial Filtering*** is selected from the ***Regression*** pull-down menu. Binomial is selected from the ***Family*** option, Pop_Hisp is inserted as the ***Dependent Variable***, and Pop_Total is inserted as the ***Normalization*** variable. The selected ***Spatial Weight Matrix*** is for the rook's definition of adjacency. Producing a pure SA description uses only the default ***intercept*** variable already appearing in the list box for independent variables. ***Eigenvalues over the principal >*** is selected, retaining its default value of 0.25, and ***Direction*** is set to ***Positive Only*** (the analysis is restricted to positive SA). Finally, the ***Show coefficients of selected eigenvectors*** and ***Save residuals and spatial filter*** boxes are checked for mapping purposes. Clicking on the ***Run*** button executes this analysis. Fig. 10.7C portrays the model output: the intercept is −1.48876, and the first three eigenvectors (EV3, EV1, and EV2, acknowledging their selection order) are selected to describe SA. Meanwhile, right-clicking on the layer opens a window from which ***Symbology*** can be selected, which in turn has a ***Choropleth*** tab (Fig. 10.7D) whose completion allows the constructed ESF to be mapped. Fig. 10.7B portrays this map.

Coding counties whose percentage of families in poverty is above the metroplex average as one, and all others as zero defines a Bernoulli RV that can be analyzed with logistic regression; this is a special binomial case in which an areal unit total count is one at most. Once more, the illustrative analysis here employs the MESF GLM specification. As before, Fig. 10.8A portrays the opened dialogue window, which requires entering the .shp file for the DFW example. Next, ***Eigenvector Spatial Filtering*** is selected from the ***Regression*** pull-down window. Logistic is selected from the ***Family*** option, and the 0–1 variable ***GTMSAPov*** is inserted as the ***Dependent Variable***. The selected ***Spatial Weight Matrix*** is for the rook's

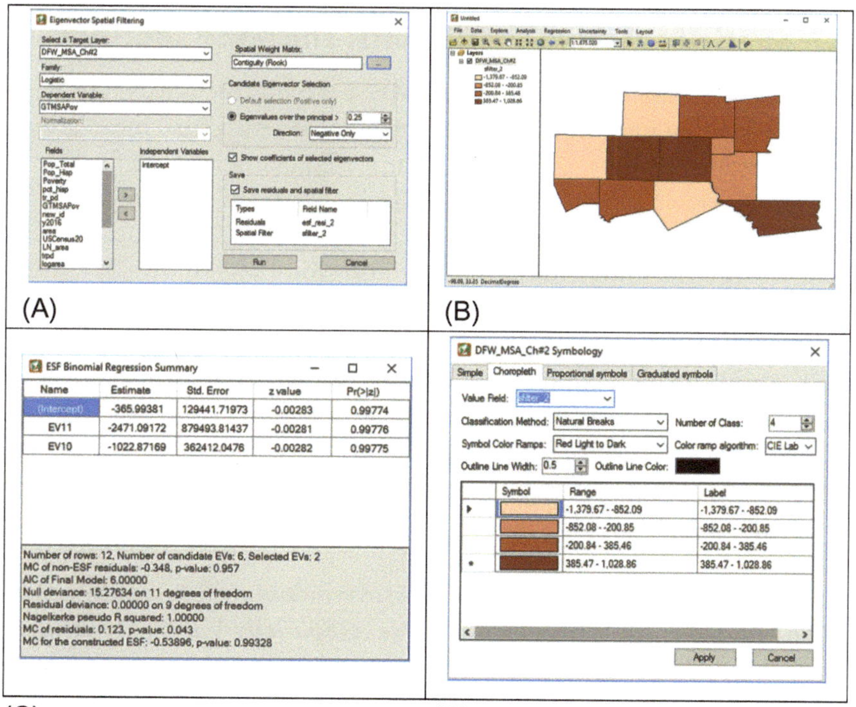

(C) (D)

Fig. 10.8 A MESF analysis of a DFW indicator variable (i.e., logistic regression) for whether a county has a poverty percentage greater than the metropolitan-wide percentage. (A) The analysis setup window. (B) The ESF map window. (C) The regression output window. (D) The choropleth mapping window.

definition of adjacency. Producing a pure SA description uses only the default ***intercept*** variable already appearing in the list box for independent variables. ***Eigenvalues over the principal >*** is selected, retaining its default value of 0.25, and ***Direction*** is set to ***Negative Only*** (because the preceding analyses use positive SA, for variety, this analysis is restricted to negative SA). Finally, the ***Show coefficients of selected eigenvectors*** and ***Save residuals and spatial filter*** boxes are checked for mapping purposes. Clicking on the ***Run*** button executes this analysis. Fig. 10.8C portrays the generated output: the intercept is −365.99381, and the tenth and eleventh eigenvectors (EV10 and EV11, ignoring their selection order) are selected to describe SA. Meanwhile, right-clicking on the layer opens a window from which ***Symbology*** can be selected, which in turn has a ***Choropleth*** tab (Fig. 10.8D) whose completion allows the constructed ESF to be mapped. Fig. 10.8B portrays this map.

10.7 Spatially varying coefficients: The MESF specification

The preceding demonstrations employ MESF to construct the equivalent of a spatially varying intercept term. This term includes a global mean around which a local mean for each areal unit varies. To better account for heterogeneity in geospatial data, this notion can be extended to regression slope coefficients (see Chapter 6, especially Sections 6.1 and 6.2). As such, selected eigenvectors (each with a mean of zero) allow local regression slope coefficients to vary around a global regression slope coefficient. Because urban economic theory states that urban neighborhoods in poverty tend to be positively correlated with population density, the indicator variable differentiating between counties with a percentage of poverty greater than the metropolitan-wide average may be included as a covariate in a Poisson regression description of population density (the log-area offset variable has a fixed coefficient of one and hence does not involve a spatially varying component).

Fig. 10.9A portrays the opened dialogue window, which requires entering the .shp file for the DFW example. Next, ***Spatially Varying Coefficients*** is selected from the ***Regression*** pull-down menu. Poisson is selected from the ***Family*** option, Pop_Total is inserted as the ***Dependent Variable***, and LN_area is inserted as the ***Offset***. The selected ***Spatial Weight Matrix*** is for the rook's definition of adjacency. To include a covariate, ***GTMSAPov*** is selected from ***Fields*** and transferred (i.e., highlight ***GTMSAPov*** and then click the > button) to ***Independent Variables***. The check mark in front of the

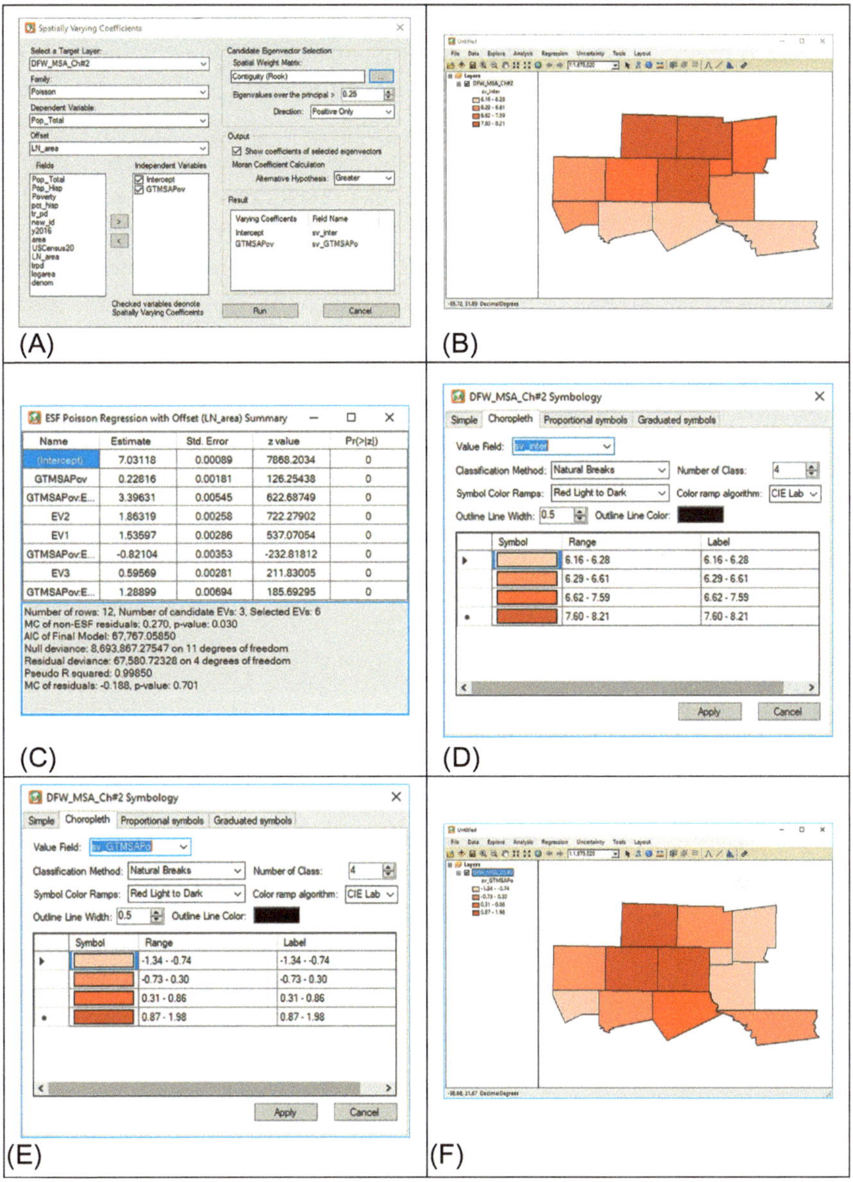

Fig. 10.9 Spatially varying coefficients constructed with MESF for population density and a single covariate. (A) The analysis setup window. (B) The spatially varying intercept ESF map window. (C) The regression output window. (D) The intercept choropleth mapping window. (E) The slope choropleth mapping window. (F) The spatially varying regression slope ESF map window.

variable names in the list box for independent variables indicates whether a covariate is specified with spatially varying coefficients. ***Eigenvalues over the principal >*** is selected, retaining its default value of 0.25. Finally, the ***Show coefficients of selected eigenvectors*** box is checked for mapping purposes; this box is checked only so that the eigenvector results are not hidden. Clicking on the ***Run*** button executes this analysis. Fig. 10.9C portrays the generated output: the intercept is 7.03118, and the first three eigenvectors (EV2, EV1, and EV3, acknowledging their order of selection) are selected for the spatially varying intercept term. Furthermore, the cross-product terms ***GTRMSAPov:EV3***, ***GTRMSAPov:EV1***, and ***GTRMSAPov:EV2*** are selected for the spatially varying slope coefficient. Meanwhile, right-clicking on the layer opens a window from which ***Symbology*** can be selected, which in turn has a ***Choropleth*** tab (Fig. 10.9D) whose completion allows the constructed spatially varying intercept term to be mapped. Fig. 10.9B portrays this map. Repeating this last step, but for the slope coefficient results (Fig. 10.9E), produces the spatially varying regression slope coefficient appearing in Fig. 10.9F.

10.8 Summary

This chapter presents output from a simple toy data example involving 12 counties forming the 2010 DFW metroplex. Analysis results reported here are not likely to be meaningful (e.g., the sample size is too small, and some reported results are not statistically significant). Demonstrations are for MESF descriptions of normal, Poisson, negative binomial, Bernoulli (0–1 logistic), and binomial RVs implemented with the SAAR software. The attribute data appear in Table 10.1. These demonstrations illustrate SAAR output for selected techniques appearing especially in Chapters 1, 3, 5, and 6.

References

Kamp, H., & Partee, B. (1995). Prototype theory and compositionality. *Cognition*, *57*, 129–191.

Koo, H., Chun, Y., & Griffith, D. (2018). Integrating spatial data analysis functionalities in a GIS environment: Spatial analysis using ArcGIS Engine and R (SAAR). *Transactions in GIS*, *22*, 721–736.

Epilogue

This book furnishes a practical overview of Moran eigenvector spatial filtering (MESF), supplementing what appears in Chun and Griffith (2013) and Griffith and Chun (2014). The Preface presents almost all of the basic data used throughout this volume. The first chapter focuses on spatial autocorrelation (SA), one of the fundamental properties of geospatial data and a phenomenon that can be found everywhere. The second chapter, perhaps the most theoretical in this book, examines spectral analysis, with special reference to georeferenced data (e.g., eigenfunctions of spatial weights matrices and satellite images) as well as MESF. It highlights both the definition of SA as map pattern and parallels between the multivariate statistics technique of principal components analysis (PCA) and MESF. It is an extension of Chapter 5 in Griffith (1988). Chapter 3 integrates MESF and linear regression, essentially translating the aforementioned PCA analogy into a fundamental theorem for MESF. It also accents the linear prediction capabilities of MESF. The fourth and tenth chapters deal with software implementations of MESF, especially with Spatial Analysis using ArcGIS Engine and R (SAAR) (Koo, Chun, & Griffith, 2018). In addition, Chapter 4 outlines the use of MESF for big spatial data. Chapter 5 integrates MESF and generalized linear model regression, essentially extending MESF to nonnormal data, such as binomial percentages and Poisson counts. The sixth chapter spotlights geographic heterogeneity in terms of both spatially varying regression coefficients and local indices of SA. Chapter 7 extends MESF to spatial interaction data analyses, Chapter 8 extends MESF to space–time data analyses, and Chapter 9 extends MESF to multivariate statistical analysis. Chapter 10 furnishes a toy example for implementation illustrations, with special reference to the SAAR software package. Overall, this book provides a quantitative spatial analyst with the tools necessary to use, efficiently and effectively, MESF to analyze virtually any georeferenced dataset (i.e., a table of attribute values linked to a .shp file).

Various implementations of MESF exist, some in R (e.g., Chun & Griffith, 2013), some in SAS, and some in other computer languages (e.g., Griffith & Chun, 2014). SAAR is a new and improving software package; it furnishes one of the few existing user-friendly implementations of MESF. It still needs to incorporate modules that execute spatial interaction and space–time analyses (Chapters 7 and 8). This software upgrading is one of a number of items constituting a future research agenda for MESF. Other agenda items include the following: fast stepwise eigenvector selection for big spatial data distributed over irregular lattices, improved error variance estimates in the presence of stepwise eigenvector selection, better integration with local SA statistics, and bolstering of prediction capabilities.

Another issue of great importance concerns repeatability and replicability of specific output. Chapter 4 raises this issue in terms of numerically calculating eigenvectors. The existence of theoretically correct vectors (Griffith, 2000) enables both repeatability and replicability in that situation. However, another source of concern is the nonlinear nature of nonnormal regression analysis. In this context, estimation solutions involve iterative routines and hence become a function of algorithm selection, rate of convergence, stopping rules, numerical precision, and initial parameter guestimates, among other considerations. For these and other reasons, random effects estimation (see Chapters 6 and 8) also can defy exact reproducibility (see Section 8.3; Baker, 2016; Goodman, Fanelli, & Ioannidis, 2016). This multifaceted context allows difficulties to arise when georeferenced data are not globally concave and computable over an entire parameter space for a MESF regression specification. One outcome could be small but detectable, or even conspicuous or glaring, differences in matched estimates. Appropriate MESF troubleshooting standards need to be devised to guide novices and other spatial analysts in these types of situations.

In conclusion, the authors hope readers of this book become avid MESF users. The authors also hope to continue their scientific collaboration to address at least some of the preceding research agenda items, and perhaps even others, and invite new and experienced MESF users to join them in these ventures.

References

Baker, M. (2016). Is there a reproducibility crisis? *Nature*, *533*, 452–454.

Chun, Y., & Griffith, D. (2013). *Spatial statistics and geostatistics*. Thousand Oaks, CA: SAGE.

Goodman, S., Fanelli, D., & Ioannidis, J. (2016). What does research reproducibility mean? *Science Translational Medicine*, *8*(341), 1–6.

Griffith, D. (1988). *Advanced spatial statistics*. Dordrecht: Kluwer.

Griffith, D. (2000). Eigenfunction properties and approximations of selected incidence matrices employed in spatial analyses. *Linear Algebra and its Applications*, *321*, 95–112.

Griffith, D., & Chun, Y. (2014). Spatial autocorrelation and spatial filtering. In M. Fischer & P. Nijkamp (Eds.), *Handbook of regional science* (pp. 1477–1507). Berlin: Springer-Verlag.

Koo, H., Chun, Y., & Griffith, D. (2018). Integrating spatial data analysis functionalities in a GIS environment: Spatial analysis using ArcGIS Engine and R (SAAR). *Transactions in GIS*, *22*, 721–736.

Index

Note: Page numbers followed by *f* indicate figures and *t* indicate tables.

MIX
Papier | Fördert gute Waldnutzung
FSC® C083411

Kolibri 360 GmbH
Ferdinand-Jühlke-Straße 7
99095 Erfurt, Deutschland
produktsicherheit@kolibri360.de